SYNCHRONIZATION IN COUPLED CHAOTIC CIRCUITS AND SYSTEMS

WORLD SCIENTIFIC SERIES ON NONLINEAR SCIENCE

Editor: Leon O. Chua
University of California, Berkeley

Series A. MONOGRAPHS AND TREATISES

Volume 25: Chaotic Dynamics in Hamiltonian Systems
H. Dankowicz

Volume 26: Visions of Nonlinear Science in the 21st Century
Edited by J. L. Huertas, W.-K. Chen & R. N. Madan

Volume 27: The Thermomechanics of Nonlinear Irreversible Behaviors — An Introduction
G. A. Maugin

Volume 28: Applied Nonlinear Dynamics & Chaos of Mechanical Systems with Discontinuities
Edited by M. Wiercigroch & B. de Kraker

Volume 29: Nonlinear & Parametric Phenomena*
V. Damgov

Volume 30: Quasi-Conservative Systems: Cycles, Resonances and Chaos
A. D. Morozov

Volume 31: CNN: A Paradigm for Complexity
L. O. Chua

Volume 32: From Order to Chaos II
L. P. Kadanoff

Volume 33: Lectures in Synergetics
V. I. Sugakov

Volume 34: Introduction to Nonlinear Dynamics*
L. Kocarev & M. P. Kennedy

Volume 35: Introduction to Control of Oscillations and Chaos
A. L. Fradkov & A. Yu. Pogromsky

Volume 36: Chaotic Mechanics in Systems with Impacts & Friction
B. Blazejczyk-Okolewska, K. Czolczynski, T. Kapitaniak & J. Wojewoda

Volume 37: Invariant Sets for Windows — Resonance Structures, Attractors, Fractals and Patterns
A. D. Morozov, T. N. Dragunov, S. A. Boykova & O. V. Malysheva

Volume 38: Nonlinear Noninteger Order Circuits & Systems — An Introduction
P. Arena, R. Caponetto, L. Fortuna & D. Porto

Volume 39: The Chaos Avant-Garde: Memories of the Early Days of Chaos Theory
Edited by Ralph Abraham & Yoshisuke Ueda

Volume 40: Advanced Topics in Nonlinear Control Systems
Edited by T. P. Leung & H. S. Qin

*Forthcoming

WORLD SCIENTIFIC SERIES ON NONLINEAR SCIENCE Series A   Vol. 41
Series Editor: Leon O. Chua

SYNCHRONIZATION IN COUPLED CHAOTIC CIRCUITS AND SYSTEMS

Chai Wah Wu
IBM Thomas J Watson Research Center, USA

World Scientific
New Jersey • London • Singapore • Hong Kong

Published by
World Scientific Publishing Co. Pte. Ltd.
P O Box 128, Farrer Road, Singapore 912805
USA office: Suite 1B, 1060 Main Street, River Edge, NJ 07661
UK office: 57 Shelton Street, Covent Garden, London WC2H 9HE

British Library Cataloguing-in-Publication Data
A catalogue record for this book is available from the British Library.

SYNCHRONIZATION IN COUPLED CHAOTIC CIRCUITS AND SYSTEMS

Copyright © 2002 by World Scientific Publishing Co. Pte. Ltd.

All rights reserved. This book, or parts thereof, may not be reproduced in any form or by any means, electronic or mechanical, including photocopying, recording or any information storage and retrieval system now known or to be invented, without written permission from the Publisher.

For photocopying of material in this volume, please pay a copying fee through the Copyright Clearance Center, Inc., 222 Rosewood Drive, Danvers, MA 01923, USA. In this case permission to photocopy is not required from the publisher.

ISBN 981-02-4713-3

Printed in Singapore by Uto-Print

To my parents

Preface

Chaos in electronic circuits can be dated back to 1927 in experiments conducted by Van Der Pol and his colleagues [1, 2]. Chaotic phenomena in engineering systems have been extensively studied and analyzed in the last few decades, but the applications have mainly been one of detection of chaos and how it can be avoided. Since the discovery by Pecora and Carroll that chaotic systems can be synchronized, the topic of synchronization of coupled chaotic circuits and systems has been investigated intensely and some interesting applications such as broadband communication systems have come out of this research. The purpose of this book is to study some aspects of synchronization of chaos in circuits and systems. Synchronization of chaos is a large research area with many topics and only a subset of these topics is discussed. In particular, the focus is on complete synchronization, where every circuit is synchronized to every other circuit, since this case is more amenable to analysis. Interesting phenomena can occur, such as clustering, wave phenomena and Turing patterns, when not all circuits are synchronized. Other related topics which are not discussed in this book include generalized synchronization and control of chaos. The reader is referred to http://www2.egr.uh.edu/~chengr/chaos-bio.html for Guanrong Chen's comprehensive list of papers in the area of synchronization and control of chaos up to 1997. A bibliography on the more general topic of chaos and nonlinear dynamics can be found in CHAOSBIB at http://www.uni-mainz.de/FB/Physik/Chaos/chaosbib.html. The prerequisite for this text is a basic knowledge of linear algebra and differential equations.

The organization of this book is as follows. After a brief introduction in

Chapter 1, synchronization in two coupled systems is discussed in Chapter 2 including applications such as communication systems. In Chapter 3, we study synchronization in arbitrarily coupled arrays of chaotic circuits and systems where the coupling elements do not have dynamics of their own. In Chapter 4 the more complicated case of dynamic coupling is considered. In Chapter 5 the relationship between the topology of the underlying graph and the synchronization criteria in arrays of oscillators is examined. So far, most of the synchronization results are obtained by means of Lyapunov's direct method. Finally, in Chapter 6 the Lyapunov exponents approach to synchronization is studied. Some background material is covered in the appendices.

I am grateful to the many people who make the writing of this book possible. I would like to thank my mentor, collaborator, colleague and friend, Prof. Leon Chua who started my interest in nonlinear dynamics and whose boundless energy and enthusiasm are an inspiration to me. I would also like to thank the following persons who I have the pleasure to meet, who I have many interesting conversations with, and who I learn a lot from: Guanrong Chen, Ken Crounse, Kevin Halle, Anshan Huang, Makoto Itoh, Michael Peter Kennedy, Ljupco Kocarev, Marco Martens, Maciej Ogorzalek, Charles Pugh, Tamás Roska, Nikolai Rulkov, Bert Shi, Mike Shub, Charles Tresser, and Guo-Qun Zhong. I also want to express my gratitude to IBM for the support I have received in writing this book. The enthusiastic support of Dr. K. K. Phua and Lakshmi Narayanan of World Scientific makes the completion of the book much easier.

I would like to thank my mother and my late father, who have shown me the virtues of hard work and integrity. And last, but not least, I would like to thank my wife Ann for her love and support, and my son Brian for all the joy he has given me.

Yorktown Heights, New York　　　　　　　　　　　　　　　　Chai Wah Wu
October 2001

Contents

Preface   vii

Chapter 1   Introduction   1

Chapter 2   Synchronization in Two Coupled Chaotic Systems   3
2.1   Pecora-Carroll subsystem decomposition   6
2.2   Separable additive coupling   6
2.3   Synchronization and stability   7
      2.3.1   Absolute stability   8
      2.3.2   Lipschitz nonlinear systems   10
      2.3.3   Circuit theoretical criteria for asymptotical stability . .   11
2.4   Communication and signal processing via synchronization of chaotic systems   13
2.5   Synchronization of nonautonomous systems   18
      2.5.1   Unidirectional synchronization scheme for nonautonomous systems   19
      2.5.2   Mutual coupling synchronization scheme for nonautonomous systems   20
      2.5.3   Synchronization between different systems   22
      2.5.4   Synchronizing nonautonomous systems as communication systems   24
2.6   Synchronization via a scalar signal   27
      2.6.1   Applications of scalar synchronization to chaotic communication system   32
2.7   Adaptive synchronization   33

	2.7.1	A general adaptive scheme	35
	2.7.2	Two coupled nonlinear systems with linear parameters	36
	2.7.3	Two coupled nonlinear systems with multiplicative parameters	40
	2.7.4	Examples	42
	2.7.5	A generalization of the scheme in Eq. (2.35)	45
	2.7.6	Adaptive observers	46
2.8		Discrete-time systems	48
2.9		Further reading	49

Chapter 3 Synchronization in Coupled Arrays of Chaotic Systems   51

3.1		Uniform linear static coupling	55
	3.1.1	G is normal	58
	3.1.2	G is symmetric	60
	3.1.3	General G	61
3.2		Uniform nonlinear static coupling	65
3.3		From stability results to synchronization criteria in coupled arrays	67
3.4		Discrete-time systems	70

Chapter 4 Synchronization in Coupled Arrays: Dynamic Coupling   77

4.1	Synchronization of clusters	81
4.2	Regular and uniform hypergraphs in linearly coupled arrays	88
4.3	Two identical systems coupled by dynamic coupling	96

Chapter 5 Graph Topology and Synchronization   99

5.1		Some coupling configurations	100
5.2		Continuous time systems	103
5.3		Discrete-time systems	107
5.4		Graph coloring via synchronized array of coupled oscillators	108
	5.4.1	Coloring two-colorable graphs	109
	5.4.2	Coloring arbitrary graphs	111
	5.4.3	Antivoter model for graph coloring	114
	5.4.4	Calculating the star chromatic number of a graph	116

Chapter 6 Lyapunov Exponents Approach to Synchronization   119
6.1 Continuous-time systems 120
6.2 Discrete-time systems 121
6.3 Three oscillator universal probe for determining synchronization in coupled arrays 122

Appendix A Some Linear Systems Theory and Matrix Theory   131

Appendix B Graph Theoretical Definitions and Notations   141

Appendix C Stability, Lyapunov's Direct Method and Lyapunov Exponents   145
C.1 Lyapunov function and Lyapunov's direct or second method . . 145
C.2 Lyapunov exponents 150

Appendix D Chaotic Circuits and Systems   153
D.1 Nonautonomous chaotic circuits and systems 153
    D.1.1 Circuit 1 153
    D.1.2 Circuit 2 156
D.2 Autonomous chaotic circuits and systems 157
    D.2.1 Chua's oscillator 157
    D.2.2 Piecewise-linear Rössler system 159
    D.2.3 Hyperchaotic electronic circuit 160
    D.2.4 Hyperchaotic Rössler system 161

Bibliography   165

Index   173

Chapter 1

Introduction

In a series of seminal papers [3, 4], Pecora and Carroll show how two chaotic systems can be synchronized by decomposing the system into two subsystems. A chaotic system $\dot{x} = f(x)$ is decomposed by the state decomposition $x = (v, u)^T$ into:

$$\dot{x} = \begin{pmatrix} \dot{v} \\ \dot{u} \end{pmatrix} = \begin{pmatrix} f_1(v, u) \\ f_2(v, u) \end{pmatrix} \quad (1.1)$$

The above equation constitutes the *drive* or *master* system. The *response* system* is derived by restricting the system to the component u, i.e. $\dot{w} = f_2(v, w)$. This decomposition is illustrated in Fig. 1.1. If the Lyapunov exponents of the system $\dot{w} = f_2(v, w)$ given a drive signal $v(t)$ from the chaotic system are all negative, then $|w(t) - u(t)| \to 0$ and we say that the response system *synchronizes* to the drive system. Since the trajectories of

---
*Also called the *driven* or *slave* system.

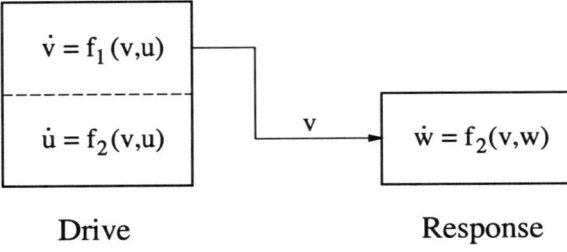

Fig. 1.1 Pecora-Carroll system decomposition.

chaotic systems are in general not periodic, the definition of synchronization used here differs from the classical definitions of synchronization used for coupled periodic oscillators. For chaotic attractors which exhibit a strong periodic component such as in the Rössler system, a generalization of the classical definition is possible [5].

Given the definition of asymptotical stability (Def. C.1), synchronization is equivalent to the fact that $\dot{w} = f_2(v(t), w)$ is asymptotically stable for all $v(t)$ where $x = (v, u)$ are trajectories of $\dot{x} = f(x)$. The main theme of this book is to explore this connection between stability and synchronization and show how stability results can be used to derive synchronization criteria of coupled nonlinear circuits and systems. We mainly focus on analytical techniques such as Lyapunov's direct method and numerical techniques such as the calculation of Lyapunov exponents. Circuit theoretical ideas are introduced whenever appropriate to illustrate their usefulness in studying synchronization in coupled circuits. For an introduction to circuit theory, the reader is referred to [6].

All the circuits and systems we consider in this book are of the form $\dot{x} = f(x, t)$ for the continuous-time case and $x(k+1) = f(x(k), k)$ for the discrete-time case. We assume that we have existence and uniqueness of trajectories for all time.

Chapter 2

Synchronization in Two Coupled Chaotic Systems

Consider two identical systems $\dot{x} = \hat{f}(x,t)$ and $\dot{y} = \hat{f}(y,t)$ as shown in Fig. 2.1. We are interested in systems $\dot{x} = \hat{f}(x,t)$ which are chaotic and

$$\boxed{\dot{x} = \hat{f}(x,t)} \qquad \boxed{\dot{y} = \hat{f}(y,t)}$$

$$\text{System 1} \qquad\qquad \text{System 2}$$

Fig. 2.1 Two identical systems $\dot{x} = \hat{f}(x,t)$ and $\dot{y} = \hat{f}(y,t)$.

exhibit sensitive dependence on initial conditions (SDIC), where nearby trajectories diverge exponentially. Since the trajectories should converge at synchronization, SDIC can be thought of as the opposite of synchronization. Thus it came as a surprise when chaotic systems can be found to synchronize. Because of SDIC, two uncoupled chaotic systems will not synchronize in general. Therefore, to achieve synchronization, we need to introduce coupling between the two systems. We assume that the two identical systems $\dot{x} = \hat{f}(x,t)$ and $\dot{y} = \hat{f}(y,t)$ are coupled in the following way:

$$\begin{aligned} \dot{x} &= f(x,x,y,t) \\ \dot{y} &= f(y,x,y,t) \end{aligned} \qquad (2.1)$$

The first argument of f is the state of the system whereas the second and third arguments of f denote the coupling from the first and the second system respectively (Fig. 2.2).

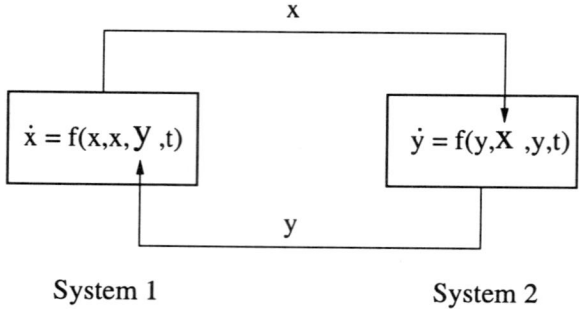

Fig. 2.2 Two coupled identical systems.

Since SDIC is a measure of instability, it is reasonable that synchronization should be related to stability. Indeed, several authors have studied the relationship between synchronization and asymptotical stability [7, 8, 9]. Furthermore, the coupling can be viewed as perturbation to the original unstable system to induce stability and several methods from control theory to stabilize an unstable system have been applied to make chaotic systems synchronizing [10, 11, 12].

The following synchronization theorem [8, 9] follows trivially from the definition of asymptotical stability (Definition C.1)*:

Theorem 2.1 *System (2.1) synchronizes in the sense that* $\|x - y\| \to 0$ *as* $t \to \infty$, *if* $\dot{x} = f(x, u(t), v(t), t)$ *is asymptotically stable for every* $u(t)$ *and* $v(t)$.

The above theorem provides a procedure on how to synchronize two identical systems $\dot{x} = \hat{f}(x,t)$ and $\dot{y} = \hat{f}(y,t)$. First rewrite $\hat{f}$ as $f(\cdot,\cdot,\cdot,t)$, i.e. $\dot{x} = f(x,x,x,t)$ and $\dot{y} = f(y,y,y,t)$ such that $\dot{x} = f(x, u(t), v(t), t)$ is asymptotically stable. Next add the coupling term $f(x,x,y,t) - f(x,x,x,t)$ to $\dot{x}$ and the coupling term $f(y,x,y,t) - f(y,y,y,t)$ to $\dot{y}$ to obtain the coupled system (2.1). Since the coupling terms are equal to 0 when $x = y$, this implies that at the synchronized state, each system exhibits the dynamics of the system $\dot{x} = f(x,x,x,t)$. Therefore each system exhibits the dynamics of the uncoupled system $\dot{x} = \hat{f}(x,t)$ at synchronization when the following

---

*All the results in this chapter (and Chapter 3) are presented as global results. Local versions can be obtained by restricting the states to some neighborhood. Note that local definitions of asymptotical stability usually require the additional condition of being Lyapunov stable. See [8, 13] for more details.

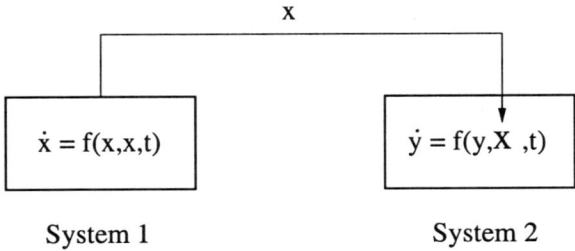

Fig. 2.3 Unidirectionally coupled systems.

consistency condition is satisfied:

$$\hat{f}(x,t) = f(x,x,x,t) \quad \text{for all } x \text{ and } t. \tag{2.2}$$

This above procedure can be roughly summarized as follows. To synchronize two chaotic systems, the parts of the system that are responsible for the instability in the system are used as the driving signal.

An important special case of coupling between two systems is the unidirectional (or master-slave) coupling in which one system does not receive any coupling from the other system (Fig. 2.3):

$$\begin{aligned} \dot{x} &= f(x,x,t) \\ \dot{y} &= f(y,x,t) \end{aligned} \tag{2.3}$$

$\dot{x} = f(x,x,t)$ is the driving or master system and $\dot{y} = f(y,x,t)$ is the response or slave system. This is the case originally studied by Pecora and Carroll and has applications in secure communication systems and chaos control [14, 15, 16].

Theorem 2.1 is very general and covers many types of synchronization between two coupled chaotic systems. It is instructive to study 2 special cases of Theorem 2.1: Pecora-Carroll decomposition and additive coupling. The reader is referred to [8] for other special cases.

2.1 Pecora-Carroll subsystem decomposition

If we define $f(x, u, z, t) = w(v(t), x)$ and $v(t)$ satisfies $\dot{v} = g(v, x)$ we obtain the Pecora-Carroll subsystem decomposition [4] :

$$\begin{aligned} \dot{v} &= g(v, x) \\ \dot{x} &= w(v, x) \\ \dot{y} &= w(v, y) \end{aligned} \qquad (2.4)$$

which would synchronize in the sense that $\|x - y\| \to 0$ as $t \to \infty$ if $\dot{x} = w(v, x)$ is asymptotically stable. The system $\dot{y} = w(v, y)$ is a reduced observer [12] for

$$\begin{aligned} \dot{v} &= g(v, x) \\ \dot{x} &= w(v, x) \end{aligned} \qquad (2.5)$$

as the state vector (v, x) can be reconstructed by using the output v and the state of the observer y.

2.2 Separable additive coupling

Consider the case where the coupling is separable and additive: $f(x, u, v, t) = \hat{f}(x, t) + K_1(x, t) + K_2(u, t) + K_3(v, t)$. In this case the coupled system (2.1) is of the form:

$$\begin{aligned} \dot{x} &= \hat{f}(x, t) + K_1(x, t) + K_2(x, t) + K_3(y, t) \\ \dot{y} &= \hat{f}(y, t) + K_1(y, t) + K_2(x, t) + K_3(y, t) \end{aligned} \qquad (2.6)$$

By Theorem 2.1 the coupled systems synchronize if

$$\dot{x} = \hat{f}(x, t) + K_1(x, t) + \eta(t) \quad \text{is asymptotically stable for all } \eta(t). \quad (2.7)$$

To satisfy consistency condition (2.2), we need $K_1 + K_2 + K_3 = 0$ in which case Eq. (2.6) can be written as

$$\begin{aligned} \dot{x} &= \hat{f}(x, t) + K_3(y, t) - K_3(x, t) \\ \dot{y} &= \hat{f}(y, t) + K_2(x, t) - K_2(y, t) \end{aligned} \qquad (2.8)$$

One interpretation of condition (2.7) is that K_1 is a *stabilizing* feedback term for $\dot{x} = \hat{f}(x, t)$ which is decomposed into two parts: $K_1 = -K_2 - K_3$. Eq. (2.8) shows how these two parts of K_1 should be coupled to ensure synchronization. When $K_3 = 0$, the coupling in Eq. (2.8) is also referred to as feedback control [16, 17].

2.3 Synchronization and stability

Theorem 2.1 relates synchronization of two coupled systems to a question of asymptotical stability. It is therefore not surprising that synchronization between two identical systems has been studied as a control theory problem and many stability results have been reformulated as synchronization criteria. For instance, in [7, 8, 18], synchronization is related to asymptotical stability and absolute stability and in [12, 11], synchronization in two systems coupled via unidirectional coupling is related to the stability of observers. In the observer problem, the question is whether the state of a system can be derived from the output of the system. For example, consider the system defined by

$$\dot{x} = f(x,t) \qquad (2.9)$$
$$y = h(x,t) \qquad (2.10)$$

where x is the state vector and y is the output vector. Then $\dot{\tilde{x}} = \tilde{f}(\tilde{x}, y, t)$ is called a full observer of Eq. (2.9) if $\tilde{x} \to x$ as $t \to \infty$, i.e. $\tilde{x}$ is a reconstruction of the state x.

For additive bidirectional coupling, there exists a similar connection between synchronization and observer design. For example, consider the following two identical systems coupled via additive bidirectional coupling

$$\dot{x} = f(x,t) + c(x,\tilde{x},t)$$
$$\dot{\tilde{x}} = f(\tilde{x},t) + \tilde{c}(x,\tilde{x},t)$$

The synchronization problem asks whether $e(t) = x - \tilde{x}$ converges to zero. Consider the corresponding observer problem:

$$\dot{y} = f(y,t)$$
$$\dot{\tilde{y}} = f(\tilde{y},t) + \tilde{c}(y,\tilde{y},t) - c(y,\tilde{y},t)$$

If $e_2(t) = y - \tilde{y}$ converges to zero, then $\tilde{y}$ is a valid observer for y. Since the dynamics of $e(t)$ and $e_2(t)$ are identical, it is clear that a valid observer implies that the two coupled systems x and $\tilde{x}$ are synchronized.

There is a difference between the observer problem and the synchronization problem. The goal in the observer problem is to reconstruct the state whereas the goal in the synchronization problem is to make the state in the response system (observer) match the state of the driving system (plant). For instance, if h is the identity function, i.e. $y = x$, then there is no need to construct an observer in order to solve the observer problem, as the state is found by observing the output directly. On the other hand, once the complete state is found, a suitable state feedback Kx will make $\dot{x} = f(x) + Kx$ asymptotically stable[†].

Consider the separable additive coupling case described in Section 2.2. To show synchronization we need to show that $\dot{x} = f(x,t) + K_1(x,t) + \eta(t) = f(x,t) - (K_2 + K_3)(x,t) + \eta(t)$ is asymptotically stable. We present several results in nonlinear control theory and nonlinear circuit theory which allow us to prove this.

2.3.1 Absolute stability

First we need the following definitions of increasing and decreasing functions [19]:

Definition 2.1 A function $\phi : R^{n+1} \to R^n$ is increasing if $(x-y)^T(\phi(x,t) - \phi(y,t)) \geq 0$ for all x, y, t.

Definition 2.2 A function $\phi : R^{n+1} \to R^n$ is strictly increasing if $(x - y)^T(\phi(x,t) - \phi(y,t)) > 0$ for all $x \neq y, t$.

Definition 2.3 A function $\phi : R^{n+1} \to R^n$ is uniformly increasing if there exists $c > 0$ such that for all x, y, t

$$(x - y)^T(\phi(x,t) - \phi(y,t)) \geq c\|x - y\|^2$$

Definition 2.4 Given a square matrix V, a function $\phi : R^{n+1} \to R^n$ is V-uniformly increasing if $V\phi$ is uniformly increasing, i.e. there exists $c > 0$ such that for all x, y, t

$$(x - y)^T V(\phi(x,t) - \phi(y,t)) \geq c\|x - y\|^2$$

A function ϕ is (V-uniformly) decreasing if $-\phi$ is (V-uniformly) increasing.

$f(x,t)$ being P-uniformly decreasing for $P = P^T > 0$ is a sufficient condition for $\dot{x} = f(x,t)$ to be asymptotically stable.

[†]under mild conditions such as Lipschitz continuity (Theorem 2.3).

Theorem 2.2 *If $f(x,t)$ is P-uniformly decreasing for a symmetric positive definite P then $\dot{x} = f(x,t) + \eta(t)$ is asymptotically stable for all $\eta(t)$.*

Proof: Construct the Lyapunov function $V = \frac{1}{2}(x-y)^T P(x-y)$. The derivative of V along trajectories of

$$\begin{aligned} \dot{x} &= f(x,t) + \eta(t) \\ \dot{y} &= f(y,t) + \eta(t) \end{aligned}$$

is

$$\dot{V} = (x-y)^T P(\dot{x} - \dot{y}) = (x-y)^T P(f(x,t) - f(y,t)) \leq -c\|x-y\|^2$$

and the conclusion follows from Theorem C.2. □

If $f(x,t)$ is differentiable in x, then $f(x,t)$ is P-uniformly decreasing if and only if $PD_1 f(x,t) + \delta I$ is negative definite for some $\delta > 0$ and all x, t. This is essentially due to the Mean Value Theorem [20].

Definition 2.5 A function $f(x,t)$ is Lipschitz continuous in x with Lipschitz constant γ if $\|f(x,t) - f(y,t)\| \leq \gamma\|x-y\|$ for all x, y, and t.

Theorem 2.3 *Suppose $f(x,t)$ is Lipschitz continuous in x with Lipschitz constant $c > 0$ and V is symmetric positive definite. If $\alpha > \frac{c}{\lambda_{\min}(V)} > 0$ where $\lambda_{\min}(V)$ is the smallest eigenvalue of V, then $\dot{x} = f(x,t) - \alpha V x$ is asymptotically stable.*

Proof: $(x-\tilde{x})^T(f(x,t) - f(\tilde{x},t) - \alpha V(x-\tilde{x}))$ is less than $(c - \alpha \lambda_{\min}(V))\|x - \tilde{x}\|^2$ by an application of Schwarz's inequality and thus $f(x,t) - \alpha V x$ is uniformly decreasing. Therefore $\dot{x} = f(x,t) - \alpha V x$ is asymptotically stable by Theorem 2.2. □

The passivity theorem states [21]:

Theorem 2.4 *Let ϕ be an increasing function and A be Hurwitz. Let $H(s)$ be defined as $C(sI - A)^{-1}B$. Suppose that (A, B, C) form a minimal realization of $H(s)$ and $H(s)$ is strictly positive real, then there exists a symmetric positive definite matrix P such that $Ax - B\phi(Cx,t) + \eta(t)$ is P-uniformly decreasing for all $\eta(t)$ and $PA + A^T P$ is negative definite. In particular, $\dot{x} = Ax - B\phi(Cx,t) + \eta(t)$ is asymptotically stable.*

For Single-Input-Single-Output systems, the above result reduces to the circle criterion [21]:

Theorem 2.5 *Let ϕ be a function such that*

$$0 \leq (y-z)(\phi(y,t) - \phi(z,t)) \leq k(y-z)^2 \qquad (2.11)$$

Let $h(s)$ be the scalar transfer function defined as $c^T(sI - A)^{-1}b$. Suppose that A is Hurwitz and (A, b, c^T) form a minimal realization of $h(s)$ and $\Re(1 + kh(j\omega)) > 0$ for all ω, then there exists a symmetric positive definite matrix P such that $Ax - b\phi(c^T x, t) + \eta(t)$ is P-uniformly decreasing for all $\eta(t)$ and $PA + A^T P$ is negative definite. In particular, $\dot{x} = Ax - b\phi(c^T x, t) + \eta(t)$ is asymptotically stable.

In [18], another criteria is given for the stability of single-input-single-output systems based on Lur'e-Postnikov type Lyapunov functions:

Theorem 2.6 *Suppose ϕ satisfies*

$$(\phi(y) - \phi(z) - k(y - z))(\phi(y) - \phi(z) - k(y - z)) \leq 0 \quad \forall y, z \quad (2.12)$$

The system $\dot{x} = Ax - b\phi(c^T x, t) - Fx + \eta(t)$ is asymptotically stable if there exists real numbers $\tau \geq 0$, $\gamma \geq 0$ and symmetric positive definite matrix P such that

$$\begin{bmatrix} (A - F)^T P + P(A - F) & \tau kc + \gamma k(A - F)^T c - Pb \\ (\tau kc + \gamma k(A - F)^T c - Pb)^T & -2(\tau + c^T b\gamma k) \end{bmatrix} < 0$$

and

$$\det \begin{bmatrix} -(A - F)^T P + P(A - F) & Pb - \tau kc \\ (Pb - \tau kc)^T & 2\tau \end{bmatrix} > 0$$

2.3.2 Lipschitz nonlinear systems

Definition 2.6 Let A be an $n \times n$ matrix and W be an $m \times n$ matrix. The pair (A, W) is called *observable* if the matrix

$$\begin{pmatrix} W \\ WA \\ WA^2 \\ \vdots \\ WA^{n-1} \end{pmatrix}$$

has rank n. The pair (A, W) is called *controllable* if the pair (A^T, W^T) is observable.

Let $\sigma_{\min}(A)$ and $\sigma_{\max}(A)$ denote the smallest and largest singular value of the matrix A respectively. In [22], the following sufficient condition for asymptotical stability of Lipschitz nonlinear systems is given:

Theorem 2.7 *The system $\dot{x} = (A - LC)x + f(x,t)$ is asymptotically stable if*

- *f is Lipschitz continuous in x for the $\|\cdot\|_2$ norm with Lipschitz constant γ,*
- *(A, C) is observable,*
- *$(A - LC)$ is Hurwitz,*
- *$\min_{\omega \geq 0} \sigma_{\min}(A - LC - j\omega I) > \gamma$.*

Furthermore, there exist symmetric $P > 0$ and $\nu > 1$ such that $(A - LC)^T P + P(A - LC) + \gamma^2 PP + \nu I = 0$.

To illustrate how stability results can be reformulated as synchronization results, Theorem 2.7 together with condition (2.7) imply that:

Theorem 2.8 *The coupled array*

$$\dot{x} = Ax + f(x,t) + L_1 C(y - x)$$
$$\dot{y} = Ay + f(y,t) + L_2 C(x - y)$$

synchronizes if

- *f is Lipschitz continuous in x for the $\|\cdot\|_2$ norm with Lipschitz constant γ,*
- *(A, C) is observable,*
- *$(A - (L_1 + L_2)C)$ is Hurwitz,*
- *$\min_{\omega \geq 0} \sigma_{\min}(A - (L_1 + L_2)C - j\omega I) > \gamma$.*

The reader is referred to [23] for more discussion on how Theorem 2.7 is useful in the synchronization between two systems.

2.3.3 *Circuit theoretical criteria for asymptotical stability*

In circuit theory, a circuit with bounded trajectories and possessing the property of asymptotical stability as defined in Definition C.1 is said to have a unique steady state. In [24] several criteria are presented which guarantee that a circuit has a unique steady state and thus also possesses asymptotical stability.

For a circuit with only two terminal elements, the following conditions imply that the circuit has a unique steady state [25, 26]:

Theorem 2.9 *A circuit has a unique steady state solution if the following conditions are met:*

(1) There is no loop in the circuit formed exclusively by capacitors, inductors and/or voltages sources.
(2) There is no cutset in the circuit formed exclusively by capacitors, inductors and/or current sources.
(3) All capacitors and inductors are passive and linear.
(4) All resistors (not including sources) are strongly locally passive.
(5) All sources are continuous and bounded.

Theorem 2.9 remains true if the capacitors and inductors are weakly nonlinear. There is a tradeoff between the local passivity of the resistors and the nonlinearity of the capacitors and inductors. The more locally passive the resistors are, the more nonlinear the capacitors and inductors can be and the theorem to still remain true [27].

We illustrate the above discussion by means of the Chua's oscillator chaotic circuit described in Appendix D. Consider two identical Chua's oscillators coupled via a linear resistor as shown in Fig. 2.4. We assume that all linear circuit elements are passive. The circuit equations are given by:

$$\begin{aligned} \frac{dv_{11}}{dt} &= \frac{1}{C_1}\left(\frac{v_{21}-v_{11}}{R} - f(v_{11}) + \frac{v_{12}-v_{11}}{R_c}\right) \\ \frac{dv_{21}}{dt} &= \frac{1}{C_2}\left(\frac{v_{11}-v_{21}}{R} + i_{31}\right) \\ \frac{di_{31}}{dt} &= -\frac{1}{L}(v_{21} + R_L i_{31}) \\ \frac{dv_{12}}{dt} &= \frac{1}{C_1}\left(\frac{v_{22}-v_{12}}{R} - f(v_{12}) + \frac{v_{11}-v_{12}}{R_c}\right) \\ \frac{dv_{22}}{dt} &= \frac{1}{C_2}\left(\frac{v_{12}-v_{22}}{R} + i_{32}\right) \\ \frac{di_{32}}{dt} &= -\frac{1}{L}(v_{22} + R_L i_{32}) \end{aligned} \qquad (2.13)$$

where $f(x) = G_b x + \frac{1}{2}(G_a - G_b)(|x+E| - |x-E|)$ is the piecewise-linear characteristic of the nonlinear resistor.

From the discussion in Section 2.2 the two Chua's oscillators synchronize if the following system of equations is asymptotically stable for all $\eta(t)$.

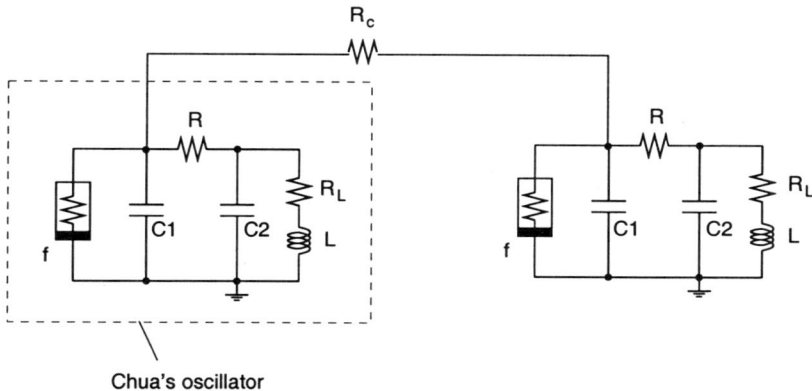

Fig. 2.4 Two identical Chua's oscillators coupled via a coupling resistor R_c.

$$\begin{aligned}
\frac{dv_1}{dt} &= \frac{1}{C_1}\left(\frac{v_2 - v_1}{R} - f(v_1) - 2\frac{v_1}{R_c} + \eta(t)\right) \\
\frac{dv_2}{dt} &= \frac{1}{C_2}\left(\frac{v_1 - v_2}{R} + i_3\right) \\
\frac{di_3}{dt} &= -\frac{1}{L}(v_2 + R_L i_3)
\end{aligned} \quad (2.14)$$

Eq. (2.14) is the state equation of Chua's oscillator driven by an external current source as shown in Fig. 2.5 where the characteristic of the nonlinear resistor is given by $g(v_1) = f(v_1) + \frac{2v_1}{R_c}$. If all the linear elements are passive $(R, L, C_1, C_2 > 0)$, and $R_c > \frac{1}{2}\max(|G_a|, |G_c|)$, then $g(v_1)$ is strongly locally passive, and Theorem 2.9 can be applied to show that Eq. (2.14) is asymptotically stable and thus the two coupled Chua's oscillators in Fig. 2.4 synchronize.

This conclusion can also be obtained by constructing a Lyapunov function directly [8].

2.4 Communication and signal processing via synchronization of chaotic systems

Since Pecora and Carroll's discovery that chaotic systems can be synchronized via unidirectional coupling, several researchers have utilized this prop-

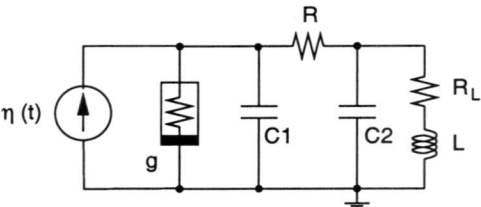

Fig. 2.5 Chua's oscillator driven by an external current source $\eta(t)$.

erty to design communication systems. In [14], a small information signal $s(t)$ is added in the transmitter to the chaotic signal $x(t)$ and transmitted. At the receiver, the transmitted signal $x(t) + s(t)$ is used to synchronize an identical chaotic system. The synchronized chaotic signal in the receiver is then used to subtract the chaos from the transmitted signal and recover the information signal. General state equations for this scheme are given by:

$$\dot{x} = f(x, x, t)$$
$$\dot{\tilde{x}} = f(\tilde{x}, x + s(t), t)$$

$c = x + s$ is the transmitted signal and the recovered information signal is given by: $r(t) = c - \tilde{x} = x - \tilde{x} + s$. This is illustrated in Figure 2.6. In this scheme, the information signal can be considered as noise to the system and the synchronization is not perfect ($\tilde{x} \neq x$), i.e. even when the transmitter and the receiver are perfectly matched with an ideal noiseless channel, there is still an inherent error $x - \tilde{x}$ in the recovered information signal. A priori, this error could be as large as or larger than the information signal. Fortunately, the error is less for the parameters used so that the recovered information signal is similar to the original information signal.

In [28, 29] this inherent error is eliminated completely by feeding the information signal back into the transmitter to generate the chaos (Fig. 2.7). For instance, consider the following chaotic communication system:

$$\dot{x} = f(x, c(x, s), t)$$
$$\dot{\tilde{x}} = f(\tilde{x}, c(x, s), t)$$

The encoding function c and the continuous decoding function d satisfy the following condition: $d(x, c(x, s)) = s$ for all x, s. For example, $c(x, s) = x + s$ and $d(\tilde{x}, w) = w - \tilde{x}$. Here s is the information signal and $c(x, s)$ is the

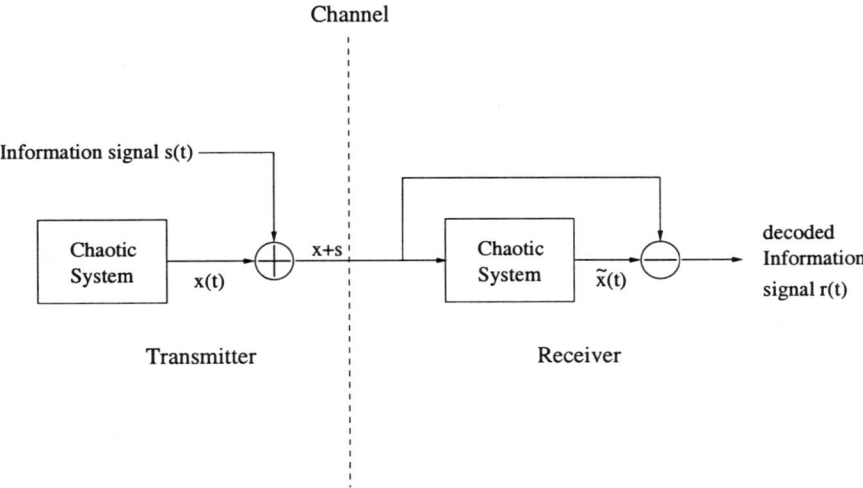

Fig. 2.6 Communication system utilizing chaos.

encoding signal that is transmitted to the receiver. At the receiver, the recovered information is constructed as $r(t) = d(\tilde{x}, c(x,s))$. If $\dot{x} = f(x, \eta(t), t)$ is asymptotically stable, then $\tilde{x} \to x$ and the recovered signal $r(t)$ will approach s. The original unperturbed chaotic system is $\dot{x} = f(x, x, t)$. If c is chosen such that $c(x, s)$ is close to x, then the transmitter system $\dot{x} = f(x, c(x, s), t)$ may remain chaotic.

In general, a communication system based on synchronized chaos can be constructed as follows:

$$\begin{aligned} \dot{x} &= f(x, s(t), t) \\ y &= h(x, s(t), t) \\ \dot{\tilde{x}} &= \tilde{f}(\tilde{x}, y, t) \end{aligned} \quad (2.15)$$

where $s(t)$ is the information signal and y is the transmitted signal (Fig. 2.8). For the system to work properly as a communication system, we need to be able to retrieve s from $\tilde{x}$. One way to achieve this is by solving the following two problems.

- First, the transmitter and the receiver need to be synchronized, i.e. $\tilde{x} \to x$ as $t \to \infty$.
- Second, use y and $\tilde{x}$ (which is an estimate of x by the first problem), to calculate s.

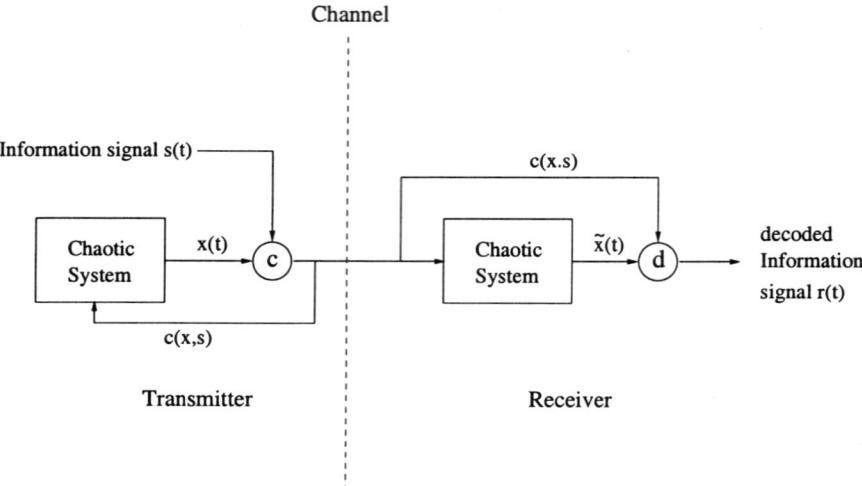

Fig. 2.7 Communication system utilizing chaos and feedback of information signal.

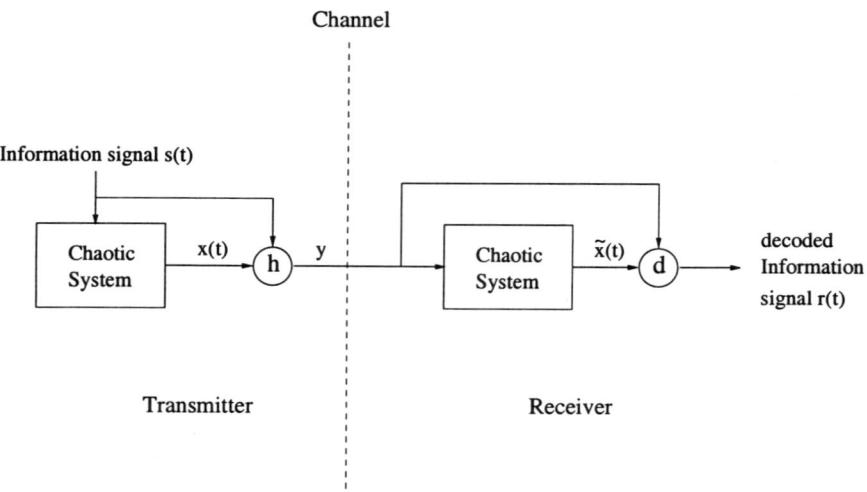

Fig. 2.8 General communication system utilizing synchronization of chaos.

For example, suppose $y = h(x, s(t), t) = x + s$. Then s can simply be derived from $s = y - x$. Another way to say this is that the transmitter in Eq. (2.15) has relative degree 0. For more complicated functions h where the relative degree is higher, another approach is needed which might not involve synchronization at all [30, 12].

As an example, consider the following communication system based on Chua's oscillator:

$$\begin{aligned}
\frac{dv_1}{dt} &= \frac{1}{C_1}[G(v_2 - v_1) - f(v_1 + r(t))] \\
\frac{dv_2}{dt} &= \frac{1}{C_2}[G(v_1 - v_2) + i_3] \\
\frac{di_3}{dt} &= -\frac{1}{L}(v_2 + R_0 i_3) \\
\frac{d\tilde{v}_1}{dt} &= \frac{1}{C_1}[G(\tilde{v}_2 - \tilde{v}_1) - f(v_1 + r(t))] \\
\frac{d\tilde{v}_2}{dt} &= \frac{1}{C_2}[G(\tilde{v}_1 - \tilde{v}_2) + \tilde{i}_3] \\
\frac{d\tilde{i}_3}{dt} &= -\frac{1}{L}(\tilde{v}_2 + R_0 \tilde{i}_3)
\end{aligned} \quad (2.16)$$

where $r(t) = \delta(a + v_1(t))(b + v_2(t))(h + i_3(t))s(t)$ and a, b, h are chosen such that $(a + v_1(t))(b + v_2(t))(h + i_3(t))$ is nonzero for all time t. The signal $v_1(t) + r(t)$ is transmitted. For $C_1, C_2, G, L > 0$ and $R_0 \geq 0$, the two systems synchronize (Section 2.3.3) and thus

$$\tilde{s}(t) = \frac{(v_1(t) + r(t)) - \tilde{v}_1(t)}{\delta(a + \tilde{v}_1(t))(b + \tilde{v}_2(t))(h + \tilde{i}_3(t))}$$

will converge to $s(t)$ as $t \to \infty$.

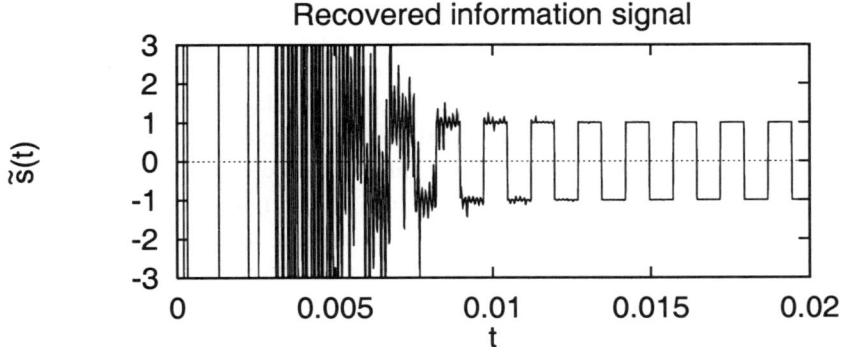

Fig. 2.9 Information signal $\tilde{s}(t)$ recovered from $v_1(t) + r(t)$. The parameters are $C_1 = 5.56nF$, $C_2 = 50nF$, $R = 1428\Omega$, $R_0 = 0\Omega$, $L = 7.14mH$, $E = 1V$, $G_a = -0.8mS$, $G_b = -0.5mS$ and $a = 2.5$, $b = 0.51$, $h = 0.0025$, $\delta = 1$.

In Fig. 2.9 we show $\tilde{s}(t)$ when $s(t)$ is a square wave of amplitude 1 and frequency $\frac{2100}{\pi}Hz$ and the parameters are chosen as $C_1 = 5.56nF$, $C_2 = 50nF$, $R = 1428\Omega$, $R_0 = 0\Omega$, $L = 7.14mH$, $E = 1V$, $G_a = -0.8mS$, $G_b = -0.5mS$ and $a = 2.5$, $b = 0.51$, $h = 0.0025$, $\delta = 1$.

Recently, there is evidence that synchronization-based chaotic communication systems proposed so far do not perform as well as incoherent chaotic communication systems where synchronization is not used [31, 32]. This is especially true in noisy environments where synchronization cannot be maintained. On the other hand, many of the incoherent chaotic communication systems proposed so far utilize the chaotic systems solely as a generator of pseudo-random signals with low cross-correlations. The rest of the communication system is similar to a traditional digital communication system [33]. Therefore it is conceivable that by exploiting knowledge about the dynamics of the systems (as synchronization-based chaotic communication systems do) a superior performance than current incoherent communication systems can be achieved. More research is needed to determine whether this is indeed the case.

2.5 Synchronization of nonautonomous systems

To apply Theorem 2.1, the two coupled systems need to be of identical form in the sense that the same f is used in both systems in Eq. (2.1). For nonautonomous or time-varying systems, where f depends on time t, this means that the time variation should be identical in the two systems. For instance, if one system is driven by a periodic signal, then the second system needs to be driven by an identical periodic signal with the same phase [34]. Generating two identical periodic signals can be difficult in practice and some phase-locking mechanism is needed [35].

By encoding all the time variation into the coupling, the time-varying part is extracted from a nonautonomous system resulting in a residual system which is autonomous. This allows two nonautonomous systems to be synchronized without the need for phase locking. We illustrate this by presenting two ways how this can be done [36].

In the first scheme, the two circuits are connected in a unidirectional driving configuration and the (not necessarily periodic) forcing is included in the driving signal so that there is no need for the response circuit to have an external forcing signal. In addition, we can recover the forcing signal at the response circuit.

In the second scheme, the two circuits are connected in a mutual coupling or bidirectional configuration. The two circuits will synchronize regardless of what the periodic forcing signals of the two circuits are. In particular, the two periodic forcing signals could have different phase, different frequency, or different shape. In fact, they two forcing signals do not have to be periodic at all.

2.5.1 Unidirectional synchronization scheme for nonautonomous systems

The chaotic system we use in this example is the nonautonomous circuit described in Appendix D and shown in Fig. D.1.

We couple two such chaotic circuits as shown in Fig. 2.10 which follows the procedure described at the beginning of this chapter. The corresponding normalized dimensionless state equations are given by:

$$\begin{align}
\frac{dx}{dt} &= k(y - f(x + A\sin(\omega t))) \\
\frac{dy}{dt} &= k\beta(-x - y) \\
\frac{d\tilde{x}}{dt} &= k(\tilde{y} - f(x + A\sin(\omega t))) \\
\frac{d\tilde{y}}{dt} &= k\beta(-\tilde{x} - \tilde{y})
\end{align} \tag{2.17}$$

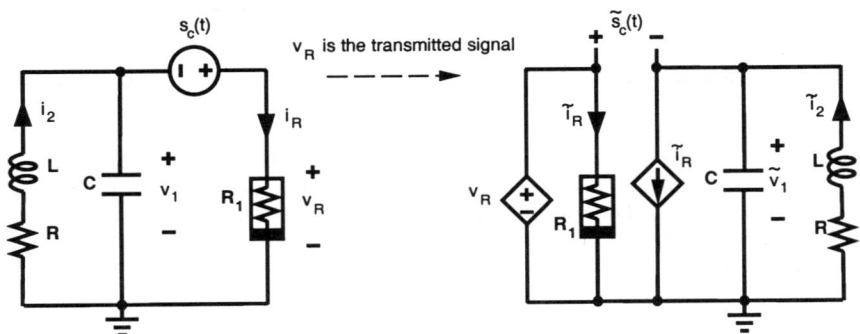

Fig. 2.10 Two nonautonomous chaotic circuits coupled through unidirectional coupling. This can be viewed as a communication system when the coupling is considered as the transmission of the signal v_R. The voltage across the controlled voltage source is v_R and the current through the controlled current source is $\tilde{i}_R$. The recovered signal $\tilde{s}_c(t)$ asymptotically approaches $s_c(t)$ as $t \to \infty$.

When $k, \beta > 0$ (corresponding to $R, L, C > 0$) this setup synchronizes, i.e. $x(t) \to \tilde{x}(t)$ as $t \to \infty$. This is easy to see since

$$\begin{pmatrix} \frac{d(x-\tilde{x})}{dt} \\ \frac{d(y-\tilde{y})}{dt} \end{pmatrix} = \begin{bmatrix} 0 & k \\ -k\beta & -k\beta \end{bmatrix} \begin{pmatrix} x - \tilde{x} \\ y - \tilde{y} \end{pmatrix} \quad (2.18)$$

which is stable as the eigenvalues have negative real parts.

The synchronized circuits imply that $\tilde{s}_c(t)$ in Fig. 2.10 approaches $s_c(t)$ as $t \to \infty$ (in the normalized equations the signal $\tilde{s}(t)$ approaches $s(t)$, where $\tilde{s}(t) = x + s(t) - \tilde{x}$).

An alternative circuit implementation of Eqs. (2.17) is to transmit the current i_R to the receiver circuit as shown in Fig. 2.11. This implementation will also synchronize the two circuits. If we assume that the nonlinear resistor R_1 is both current and voltage controlled (i.e., $f_c(v)$ is a bijection), then $\tilde{s}_c(t)$ will approach $s_c(t)$ as $t \to \infty$.

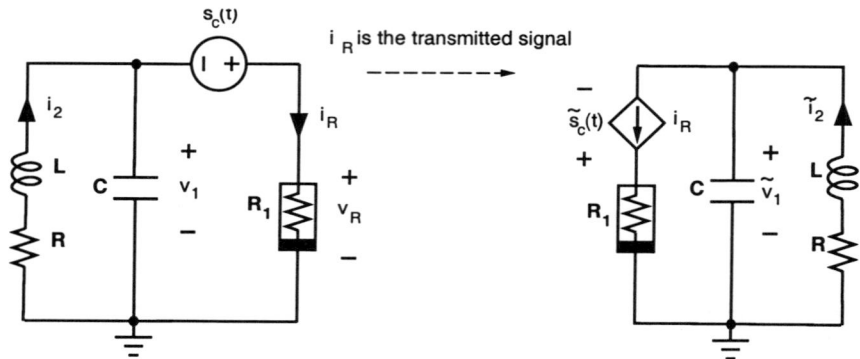

Fig. 2.11 Alternative way of connecting two nonautonomous chaotic circuits through unidirectional coupling. This can be viewed as a communication system when the coupling is considered as the transmission of the signal i_R. The current through the controlled current source is i_R. The recovered signal $\tilde{s}_c(t)$ will asymptotically approach $s_c(t)$ as $t \to \infty$ assuming that R_1 is both voltage and current controlled.

2.5.2 Mutual coupling synchronization scheme for nonautonomous systems

The chaotic system we use in this example is the nonautonomous circuit described in Appendix D and shown in Fig. D.4.

We synchronize two such circuits by connecting a linear resistor of resistance R_c across the two linear resistors, as shown in Fig. 2.12. For this case, the circuit parameters are chosen such that $R < 0$ is active, and the nonlinear resistor R_1 is passive.

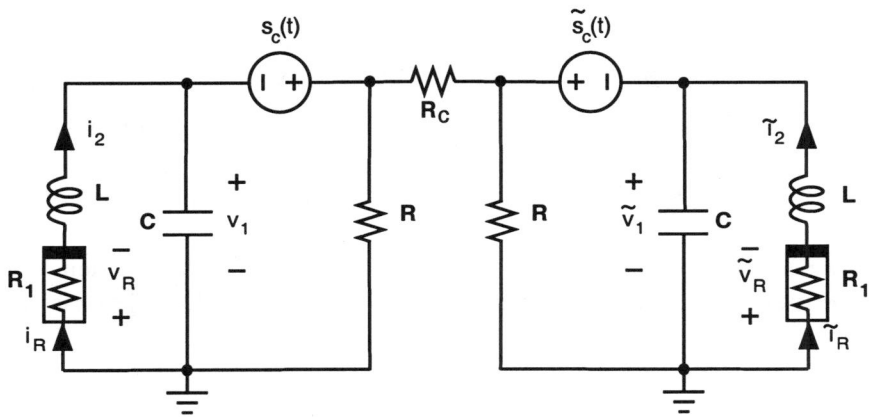

Fig. 2.12 Two nonautonomous chaotic circuits coupled through a linear resistor. The signal $\tilde{v}_1(t)$ approaches $v_1(t)$ as the system synchronizes.

The normalized state equations of the system in Fig. 2.12 are:

$$\begin{aligned}
\frac{dx}{dt} &= k(y - x - s(t) + \gamma(\tilde{x} + \tilde{s}(t) - x - s(t))) \\
\frac{dy}{dt} &= k\beta(-x - f(y)) \\
\frac{d\tilde{x}}{dt} &= k(\tilde{y} - \tilde{x} - \tilde{s}(t) + \gamma(x + s(t) - \tilde{x} - \tilde{s}(t))) \\
\frac{d\tilde{y}}{dt} &= k\beta(-\tilde{x} - f(\tilde{y}))
\end{aligned} \quad (2.19)$$

where $\gamma = \frac{1}{GR_c}$.

Let us assume that R_1 is strictly passive, i.e., $\mu = \sup_{y \neq y'} \frac{f(y) - f(y')}{y - y'} < 0$. With the chosen circuit parameters, this implies that $\beta > 0$ and $k < 0$. We choose $\gamma = -\frac{1}{2}$. This corresponds to $R_c = -2R$. Since $R < 0$ is active, this implies that $R_c > 0$ is a passive resistor. Eqs. (2.19) can then be rewritten as

$$\begin{aligned}
\frac{dx}{dt} &= k(y - [\tfrac{1}{2}(x + \tilde{x}) + \tfrac{1}{2}(s(t) + \tilde{s}(t))]) \\
\frac{dy}{dt} &= k\beta(-x - f(y)) \\
\frac{d\tilde{x}}{dt} &= k(\tilde{y} - [\tfrac{1}{2}(x + \tilde{x}) + \tfrac{1}{2}(s(t) + \tilde{s}(t))]) \\
\frac{d\tilde{y}}{dt} &= k\beta(-\tilde{x} - f(\tilde{y}))
\end{aligned} \quad (2.20)$$

If we set $\eta(t) = -\frac{1}{2}(x + \tilde{x}) - \frac{1}{2}(s(t) + \tilde{s}(t))$, then by the discussion in Section 2.2, this system synchronizes (i.e. $\tilde{x} \to x$ and $\tilde{y} \to y$ as $t \to \infty$) if the following system is asymptotically stable for all $\eta(t)$.

$$\begin{pmatrix} \frac{dx}{dt} \\ \frac{dy}{dt} \end{pmatrix} = k \begin{pmatrix} y \\ \beta(-x - f(y)) \end{pmatrix} + \begin{pmatrix} k\eta(t) \\ 0 \end{pmatrix} \tag{2.21}$$

We will use Lyapunov's direct method and the following Lyapunov function:

$$V = \begin{pmatrix} x - \tilde{x} \\ y - \tilde{y} \end{pmatrix}^T \begin{pmatrix} 1 & -c \\ -c & \frac{1}{\beta} \end{pmatrix} \begin{pmatrix} x - \tilde{x} \\ y - \tilde{y} \end{pmatrix}$$

If we choose c to be small and positive, then the Lyapunov function V is positive definite. Taking the derivative of V with respect to trajectories we get:

$$\begin{aligned}
\dot{V} &= kc\beta(x - x')^2 + kc\beta(x - x')(f(y) - f(y')) - kc(y - y)^2 \\
&\quad - k(y - y')(f(y) - f(y')) \\
&= k\left(c\beta(x - x')^2 + c\beta s(x - x')(y - y') + (-c - s)(y - y')^2\right)
\end{aligned}$$

where $s = \frac{f(y) - f(y')}{y - y'} \leq \mu < 0$. This simplifies to:

$$k\left(\sqrt{c\beta}(x - x') + \frac{s\sqrt{c\beta}}{2}(y - y')\right)^2 + k\left(\frac{-c\beta s^2}{4} - c - s\right)(y - y')^2$$

For sufficiently small $c > 0$, $\frac{-c\beta s^2}{4} - c - s > -\mu$ and thus $\dot{V} < -k\mu(y - y')^2$. Therefore system (2.21) is asymptotically stable for all $\eta(t)$ by Theorem C.2 and system (2.20) synchronizes for all $s(t)$ and $\tilde{s}(t)$.

Note that when the two systems are synchronized ($x = \tilde{x}$ and $y = \tilde{y}$), each of the two systems behaves as a single system (D.7) with the external source replaced by the average of the two sources $s_c(t)$ and $\tilde{s}_c(t)$.

2.5.3 Synchronization between different systems

Since the external driving forces in the two systems in Section 2.5.2 are arbitrary, the synchronization can be considered as synchronization between two different systems. For instance, we can add a nonlinear (resistive or dynamic) one-port in series with the external forcing voltage source $S_c(t)$ and the coupled system will still synchronize. In this case the two systems

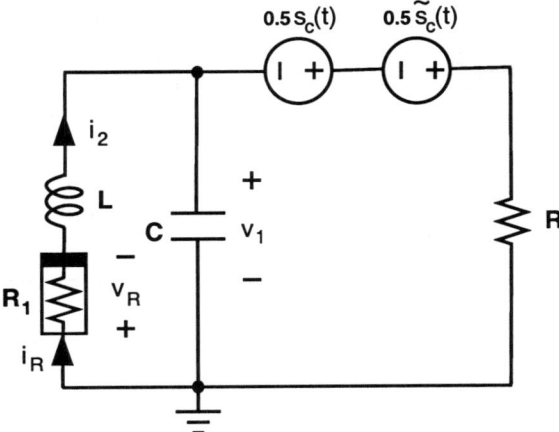

Fig. 2.13 At synchronization, each of the two systems in Fig. 2.12 behaves as a single system with the external source consisting of the average of the two sources $s_c(t)$ and $\tilde{s}_c(t)$.

are not identical and can have different dimensions. Synchronization here means that the state variables in one system which have a corresponding counterpart in the other system will approach each other as $t \to \infty$ (see Definition 8 in [8]).

For the unidirectional configuration (Fig. 2.10) this leads to Fig. 2.14, where the added nonlinear one-port is shown as N_2. For example, assume that the one-port N_2 consists of a linear capacitor in series with a nonlinear resistor. The resulting system is shown in Fig. 2.15. We assume that R_1 is voltage controlled and the driving point characteristic of R_1 in series with R_2 is also voltage controlled. Then $\tilde{v}_1 \to v_1$ and $\tilde{i}_2 \to i_2$ as $t \to \infty$.

When the one-port N_2 is dynamic, this synchronization scheme is similar to the homogeneous driving scheme of Pecora and Carroll where the response system is a lower dimensional system than the driving system (Section 2.1).

Similarly, two one-ports can be connected in series to the independent sources in Fig. 2.12 and the system will still synchronize[‡] in the sense that $\tilde{v}_1 \to v_1$ and $\tilde{i}_2 \to i_2$ as $t \to \infty$ (Fig. 2.16).

These approaches can also be used to synchronize two different au-

---

[‡]As long as we can write state equations for the entire system.

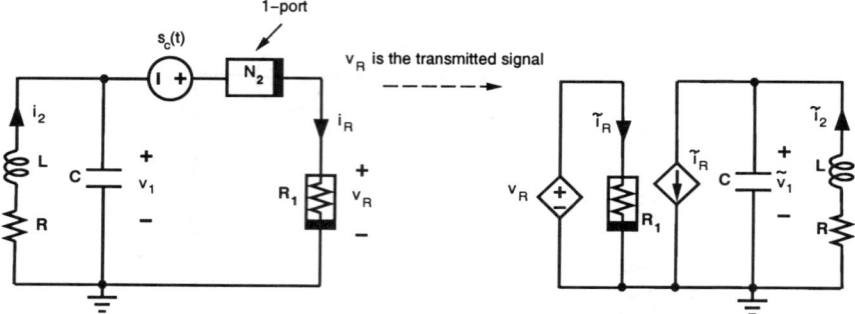

Fig. 2.14 Synchronization of two different systems. The one-port N_2 can be dynamic or resistive.

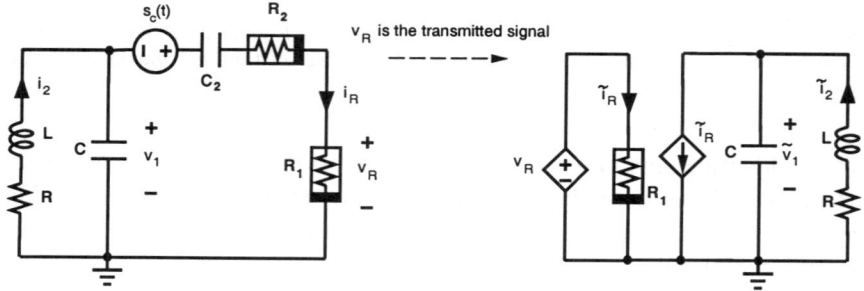

Fig. 2.15 Figure 2.14 redrawn when the one-port N_2 is a linear capacitor in series with a nonlinear resistor. $\tilde{v}_1 \to v_1$ and $\tilde{i}_2 \to i_2$ as $t \to \infty$.

tonomous systems with a similar topology, for instance Chua's oscillator (Appendix D). The reader is referred to [36] for more details including experimental results.

2.5.4 *Synchronizing nonautonomous systems as communication systems*

We have shown how two nonautonomous chaotic systems can be synchronized without the need to explicitly phase-lock the forcing in the two systems. In these synchronization schemes the external forcing of the two systems can be arbitrary and do not need to be identical. This allows us to interpret them as communication systems and synchronization schemes

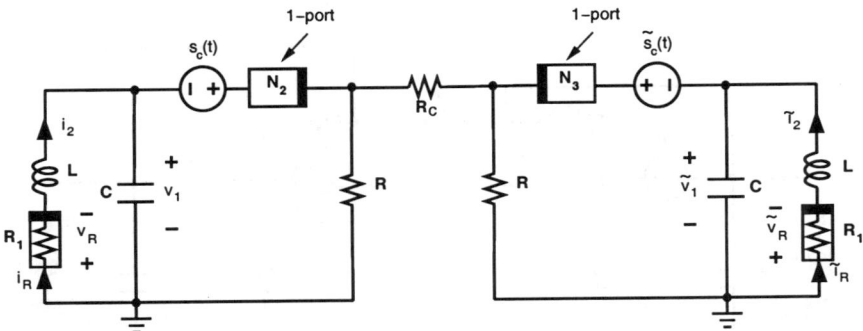

Fig. 2.16 Figure 2.12 with two one-ports N_2 and N_3 inserted in series with the external sources. The system synchronizes in the sense that $\tilde{v}_1 \to v_1$ and $\tilde{i}_2 \to i_2$ as $t \to \infty$.

for two different systems.

In particular, in the unidirectional configuration (Section 2.5.1) the response system does not need an external periodic forcing and can recover the periodic forcing of the drive system. This suggests the possibility of using this as a communication system by replacing the external periodic forcing signal in the drive circuit by a properly encoded information signal, which can then be recovered in the response circuit.

In the mutual coupling scheme (Section 2.5.2), the two periodic forcing signal can be completely different. This suggests the possibility to use this as a bidirectional communication system with the two systems both receiving and transmitting at the same time. Both schemes can also be considered as synchronization schemes for two systems which are not identical.

There have been many approaches to implementing communication systems based on chaotic synchronization [29, 37, 14, 15, 38, 39, 40, 41]. In these systems, the input signal is scrambled or converted to a chaotic signal in the transmitter and this chaotic signal is transmitted to the receiver. Nearly all of them utilize an autonomous chaotic system and the information signal does not play a significant role in generating the chaos.

Let us now consider the above synchronization schemes in this light. The synchronization schemes in Sections 2.5.1 and 2.5.2 can be considered as communication systems if the periodic signal $s_c(t)$ in Fig. 2.10 (resp. signals $s_c(t)$ and $\tilde{s}_c(t)$ in Fig. 2.12) is replaced by an encoded information signal that oscillates at the proper rate. Some examples of encoding of binary information signals that could be used are coded PCM (Manchester

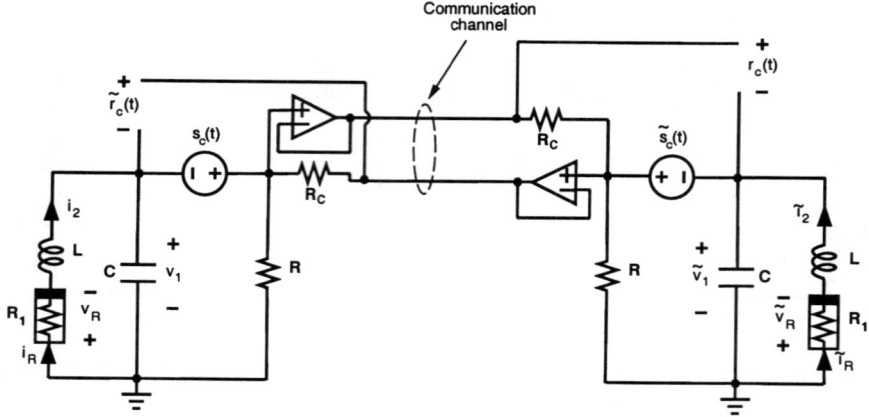

Fig. 2.17 The system in Fig. 2.12 redrawn as a bidirectional communication system. The signal $s_c(t)$ is recovered in the second system as $r_c(t)$ ($r_c(t) \to s_c(t)$ as $t \to \infty$) and $\tilde{s}_c(t)$ is recovered in the first system as $\tilde{r}_c(t)$ ($\tilde{r}_c(t) \to \tilde{s}_c(t)$ as $t \to \infty$).

pulses), FSK or PSK. For FSK and PSK, the frequencies of the keys should be chosen such that system (D.1) (resp. system (D.5)) is chaotic.

Consider Figure 2.10. The information signal is $s_c(t)$, and this is scrambled by the circuit. The scrambled signal v_R is then transmitted, and in the receiver the signal $\tilde{s}_c(t)$ is recovered which approaches $s_c(t)$. The signal $\tilde{s}_c(t)$ is now an information-bearing signal at the response circuit (receiver). Compared with the communication schemes using autonomous chaotic systems discussed above, we note that:

- The information signal plays a crucial role in generating the chaotic signal to be transmitted.
- The minimum number of dimensions needed to generate chaos is less (2 versus 3).

The scheme in Fig. 2.12 can be redrawn as a bidirectional communication system, as shown in Fig. 2.17. Both circuits transmit and receive to each other at the same time. The signals $s_c(t)$ and $\tilde{s}_c(t)$ are both information signals. The signal $s_c(t)$ is recovered in the second system as $r_c(t)$ ($r_c(t) \to s_c(t)$ as $t \to \infty$) and $\tilde{s}_c(t)$ is recovered in the first system as $\tilde{r}_c(t)$ ($\tilde{r}_c(t) \to \tilde{s}_c(t)$ as $t \to \infty$).

2.6 Synchronization via a scalar signal

In unidirectional coupling, when one chaotic system is considered as a transmitter and the other as a receiver, the entire coupled system can be viewed as a communication system (Section 2.4). In these applications, it is desirable that the two chaotic systems can be synchronized via a scalar signal since this minimizes the number of signals that need to be transmitted.

At first it was believed that the number of scalar signals needed to synchronize two chaotic systems is related to the number of positive Lyapunov exponents. In particular, it was believed that hyperchaotic systems, i.e., systems having more than one positive Lyapunov exponents, cannot be synchronized by a scalar signal. However, it has since been shown numerically and analytically that this is false. In [42, 43] it was shown numerically that hyperchaotic nonlinear dynamical systems can be synchronized with a scalar signal. In [44] it was shown how the full state of the drive system can be derived from a scalar signal via a reconstruction technique. Once the complete state of the drive system is known the response system can be synchronized to the drive system using full state feedback.

Another class of techniques to synchronize two chaotic systems via a scalar signal is based on the well-known problem of eigenvalue assignment. In [12, 11] the eigenvalue assignment problem is approached by means of observer design in linear output feedback systems. Using these techniques, we present in this section two classes of nonlinear dynamical systems which can be proved to be synchronizable via a scalar signal. The classes include several systems in the literature believed to be hyperchaotic on the basis of numerical experiments.

The main idea is that by choosing an appropriate output feedback in observable systems, the closed loop system is linear and the eigenvalues of the closed loop system can be arbitrary set and thus the rate of convergence to the equilibrium point can be arbitrarily set. By choosing analogously the proper signal to be transmitted, the rate of convergence to the synchronized state can be arbitrarily set. We illustrate these results using two hyperchaotic systems studied in the literature. We show how this can be used in chaotic signal scrambling.

Consider the drive system:

$$\dot{x} = f(x) \tag{2.22}$$

where $x \in R^n$. From the state trajectory $x(t)$ we can construct a scalar signal $y(t)$ via a function $q: R^n \to R$:

$$y = q(x).$$

By coupling y into an identical response system $\dot{\tilde{x}} = f(\tilde{x})$, we want the response system to synchronize to the drive system, i.e., $\tilde{x} \to x$ as $t \to \infty$. We will consider the case where y is coupled additively into the response system, i.e.,

$$\dot{\tilde{x}} = f(\tilde{x}) + g(y, \tilde{x}). \qquad (2.23)$$

If the functions q and g can be found such that the response system synchronizes to the drive system, i.e., trajectories $\tilde{x}$ of Eq. (2.23) converge towards trajectories of x of Eq. (2.22) as $t \to \infty$, then we say that f is *synchronizable*. Notice that the synchronization is enforced through the coupling of a scalar variable y. The following two classes of nonlinear systems are synchronizable for almost every member in the class.

Definition 2.7 A dynamical system belongs to class $\mathcal{K}_1$ if it is of the form

$$\dot{x} = Ax + f_1(w^T x) \qquad (2.24)$$

where x and w are n-vectors, A is an $n \times n$ matrix and f_1 is a function from R into R^n.

An example of a system in class $\mathcal{K}_1$ is given by Eq. (D.11).

Definition 2.8 A dynamical system belongs to class $\mathcal{K}_2$ if it is of the form

$$\dot{x} = Ax + bh(x) + d \qquad (2.25)$$

where x, b and d are n-vectors, A is an $n \times n$ matrix and h is a scalar-valued function from R^n into R.

An example of a system in class $\mathcal{K}_2$ is given by Eq. (D.12). When restricted to the scalar input or scalar output case, the definitions of observability and controllability (Definition 2.6) reduce to:

Definition 2.9 Let A be an $n \times n$ matrix and w be an n-vector. The pair (A, w^T) is called *observable* if the matrix

$$K = K(A, w^T) = \begin{pmatrix} w^T \\ w^T A \\ w^T A^2 \\ \vdots \\ w^T A^{n-1} \end{pmatrix}$$

is nonsingular. The pair (A, w) is called *controllable* if the pair (A^T, w^T) is observable.

The following theorem gives a solution to the eigenvalue assignment problem encountered in single-input linear feedback systems:

Theorem 2.10 *Let A be a $n \times n$ matrix and b be a n-vector. If (A, b) is controllable, and p is an n-th order monic polynomial with real coefficients, then there exists an n-vector w such that the characteristic polynomial of $A + bw^T$ is p.*

An algorithm for finding w is given in [45, page 337]. By duality, the following result is also true:

Corollary 2.1 *Let A be a $n \times n$ matrix and w be an n-vector. If (A, w^T) is observable, and p is an n-th order monic polynomial with real coefficients, then there exists a n-vector b such that the characteristic polynomial of $A + bw^T$ is p.*

The following two theorems give conditions under which systems can be made synchronizable by coupling a scalar signal.

Theorem 2.11 *Consider Eq. (2.24). If (A, w^T) is observable, then a system in class $\mathcal{K}_1$ is synchronizable via a scalar signal.*

Proof: Choose a polynomial p such that all its roots are in the open left half plane. By Corollary 2.1, we can find b such that the characteristic polynomial of $A + bw^T$ is p.

We synchronize two systems in $\mathcal{K}_1$ as follows:

$$\begin{aligned} \dot{x} &= Ax + f_1(w^T x) \quad \text{[drive system]} \\ \dot{\tilde{x}} &= A\tilde{x} + f_1(w^T x) - bw^T(x - \tilde{x}) \quad \text{[response system]} \end{aligned} \quad (2.26)$$

The scalar signal $w^T x$ is transmitted from the drive system to the response system. The difference in the state variables $x - \tilde{x}$ satisfies:

$$\dot{x} - \dot{\tilde{x}} = (A + bw^T)(x - \tilde{x}). \tag{2.27}$$

Since all the eigenvalues of $A + bw^T$ are in the open left half plane, $x - \tilde{x} \to 0$ as $t \to \infty$. $\square$

The dual result is:

Theorem 2.12 *Consider Eq. (2.25). If (A, b) is controllable, then a system in class $\mathcal{K}_2$ is synchronizable via a scalar signal.*

Proof: Choose a polynomial p such that all its roots are in the open left half plane. By Theorem 2.10, we can find w such that the characteristic polynomial of $A + bw^T$ is p.

We synchronize two systems in $\mathcal{K}_2$ as follows:

$$\begin{aligned} \dot{x} &= Ax + \underbrace{bh(x)}_{\text{drive system}} + d \\ \dot{\tilde{x}} &= A\tilde{x} + \underbrace{b(h(x) - w^T x)}_{\text{response system}} + bw^T \tilde{x} + d \end{aligned} \tag{2.28}$$

The scalar signal $h(x) - w^T x$ is transmitted from the drive system to the response system. The difference in the state variables $x - \tilde{x}$ again satisfies Eq. (2.27). Since all the eigenvalues of $A + bw^T$ are in the open left half plane, $x - \tilde{x} \to 0$ as $t \to \infty$. $\square$

The proofs of Theorems 2.11 and 2.12 give explicit algorithms for determining the scalar signal which synchronizes the two coupled systems. Since the polynomial p can be chosen arbitrarily, the rate of convergence to the synchronized state can be set arbitrarily.

To illustrate these results, we choose an example of a hyperchaotic dynamical system from each of the two classes $\mathcal{K}_1$ and $\mathcal{K}_2$ which we synchronize via Theorems 2.11 and 2.12 respectively.

The state equations of the hyperchaotic electronic circuit (Eq. (D.11)) in Section D.2.3 belong to class $\mathcal{K}_1$. By choosing p to be $p(\lambda) = (\lambda + 1)^4$ and applying Theorem 2.11, we can synchronize two such circuits by transmitting the scalar signal $v_2 - v_1$ as follows:

$$\frac{dv_1}{dt} = \frac{1}{C_1}(f(v_2 - v_1) - i_1)$$
$$\frac{dv_2}{dt} = \frac{1}{C_2}(-f(v_2 - v_1) - i_2)$$
$$\frac{di_1}{dt} = \frac{1}{L_1}(v_1 + Ri_1)$$
$$\frac{di_2}{dt} = \frac{1}{L_2}v_2$$
$$\frac{d\tilde{v}_1}{dt} = \frac{1}{C_1}\left(f(v_2 - v_1) - \tilde{i}_1\right) - 0.1241(v_2 - v_1 - \tilde{v}_2 + \tilde{v}_1)$$
$$\frac{d\tilde{v}_2}{dt} = \frac{1}{C_2}(-f(v_2 - v_1) - \tilde{i}_2) + 4.8759(v_2 - v_1 - \tilde{v}_2 + \tilde{v}_1)$$
$$\frac{d\tilde{i}_1}{dt} = \frac{1}{L_1}(\tilde{v}_1 + R\tilde{i}_1) - 0.2371(v_2 - v_1 - \tilde{v}_2 + \tilde{v}_1)$$
$$\frac{d\tilde{i}_2}{dt} = \frac{1}{L_2}\tilde{v}_2 + 1.0263(v_2 - v_1 - \tilde{v}_2 + \tilde{v}_1)$$

The state equations of the hyperchaotic Rössler system (Eq. (D.12)) in Section D.2.4 belong to class $\mathcal{K}_2$. By choosing the same polynomial p as before and applying Theorem 2.12, we can synchronize two such circuits by transmitting the scalar signal $xz - 3.3712x - 0.9561y + 4.3z - 5.8126w$ as follows:

$$\frac{dx}{dt} = -y - z$$
$$\frac{dy}{dt} = x + 0.25y + w$$
$$\frac{dz}{dt} = 3 + xz$$
$$\frac{dw}{dt} = -0.5z + 0.05w$$
$$\frac{d\tilde{x}}{dt} = -\tilde{y} - \tilde{z}$$
$$\frac{d\tilde{y}}{dt} = \tilde{x} + 0.25\tilde{y} + \tilde{w}$$
$$\frac{d\tilde{z}}{dt} = 3 + xz - 3.3712x - 0.9561y + 4.3z - 5.8126w$$
$$\qquad -(-3.3712\tilde{x} - 0.9561\tilde{y} + 4.3\tilde{z} - 5.8126\tilde{w})$$
$$\frac{d\tilde{w}}{dt} = -0.5\tilde{z} + 0.05\tilde{w}$$

For a given A and p, there are many choices for w and b such that $A + bw^T$ has p as the characteristic polynomial. For example, in [46] it is shown that dissipative linear systems with distinct eigenvalues can be synchronized using only a scalar signal, regardless of the eigenvalue pattern. In particular, a different algorithm than [45] is given for generating the vectors b and w to set the eigenvalues of $A + bw^T$. In fact, the results in [46] can be used to prove a slightly weaker version of Theorem 2.10:

Theorem 2.13 *Let A be an $n \times n$ matrix with distinct eigenvalues and p be an n-th order monic polynomial with real roots at least $n-1$ of which are distinct. Then there exists n-vectors b and w such that the characteristic polynomial of $A + bw^T$ is p.*

Proof: Let $\{d_i\}$ be the roots of p such that $d_2, \ldots, d_n$ are distinct. According to [46], there exists P such that PAP^{-1} is of the form

$$\begin{pmatrix} a & c_2 & \cdots & c_n \\ b_2 & d_2 & & 0 \\ \vdots & & \ddots & \\ b_n & 0 & & d_n \end{pmatrix}$$

where $a = \text{trace}(A) - \sum_{i=2}^{n} d_i$. It is then clear that $PAP^{-1} + bw^T$ has characteristic polynomial p where $b = (b_1, \ldots, b_n)^T$, $w = (-1, 0, \ldots)^T$ and $b_1 = a - d_1 = \text{trace}(A) - \sum_{i=1}^{n} d_i$. □

Note that Theorem 2.10 implies Theorem 2.13. This follows from the fact that if A is a diagonal matrix with distinct eigenvalues, then the matrix $K(A, (1, \ldots, 1))$ is a nonsingular Vandemonde matrix and thus $(A, (1, \ldots, 1))$ is observable.

2.6.1 Applications of scalar synchronization to chaotic communication system

As mentioned before, in communication systems, we would like to minimize the number of signals that are sent. Therefore being able to synchronize with a scalar signal is beneficial to the design of chaotic communication systems. In this section we briefly describe how the unidirectional synchronization of two nonlinear dynamical systems described above can be used to implement chaotic signal scrambling systems. As in Section 2.4, we need a pair of scalar-valued encoding-decoding functions c and d such that $d(x, c(x, s)) = s$ for all x and s and d is continuous.

For dynamical systems in class $\mathcal{K}_1$, Eq. (2.26) is modified into:

$$\begin{aligned} \dot{x} &= Ax + f_1(c(x,s)) + b(w^T x - c(x,s)) \qquad \text{[drive system]} \\ \dot{\tilde{x}} &= A\tilde{x} + f_1(c(x,s)) + b(w^T \tilde{x} - c(x,s)) \qquad \text{[response system]} \end{aligned} \qquad (2.29)$$

The signal $s(t)$ is the original information signal. The scrambled scalar signal $c(x(t), s(t))$ is transmitted to the response system (receiver). To maintain that the drive system here generates trajectories similar to the original drive system, we require that $c(x, s) \approx w^T x$. The difference $x - \tilde{x}$ will again satisfy Eq. (2.27) and thus $\tilde{x} \to x$ as $t \to \infty$. We can then use the decoding function d in the response system to recover the information signal since $d(\tilde{x}, c(x, s(t))) \to s(t)$ as $t \to \infty$ by the continuity of d.

Similarly, for dynamical systems in class $\mathcal{K}_2$, the following equations are used:

$$\begin{aligned}\dot{x} &= Ax + b(\underbrace{c(x,s) + w^T x}_{\text{drive system}}) + d \\ \dot{\tilde{x}} &= A\tilde{x} + b(\underbrace{c(x,s) + w^T \tilde{x}}_{\text{response system}}) + d\end{aligned} \qquad (2.30)$$

where $c(x, s)$ is the transmitted signal. To maintain that the drive system here generates trajectories similar to the original drive system, we require that $c(x, s) \approx h(x) - w^T x$.

2.7 Adaptive synchronization

So far, we have mostly considered coupling of two identical systems. For example, the circuit parameters in the two coupled Chua's oscillators in Section 2.3.3 are matched in order for synchronization to occur. When the parameters are mismatched, as is the case in physical circuits, synchronization error can occur.

In this section, we present synchronization where the parameters of the two systems are modified by an adaptive algorithm with the goal to synchronize the two systems. In some, but not in all cases, this implies that the parameters of the two systems approach each other. Some work in this area include [47, 48, 49, 50, 51, 52]. In particular, we discuss a variation of the adaptive scheme in [47] which is more amenable to analysis. We prove that for nonlinear systems with linear parameters, the synchronization error approaches zero.

Consider two coupled continuous-time systems described by:

$$\left.\begin{aligned}\dot{x}_1 &= f_1(x, a_{11}, \ldots, a_{1m}, x, \tilde{x}, t) \\ &\vdots \\ \dot{x}_n &= f_n(x, a_{n1}, \ldots, a_{nm}, x, \tilde{x}, t)\end{aligned}\right\} \leftarrow \boxed{\text{System 1}} \qquad (2.31)$$

$$\left. \begin{aligned} \dot{\tilde{x}}_1 &= f_1(\tilde{x}, \tilde{a}_{11}, \ldots, \tilde{a}_{1m}, x, \tilde{x}, t) \\ &\vdots \\ \dot{\tilde{x}}_n &= f_n(\tilde{x}, \tilde{a}_{n1}, \ldots, \tilde{a}_{nm}, x, \tilde{x}, t) \end{aligned} \right\} \leftarrow \boxed{\text{System 2}} \qquad (2.32)$$

where $x = (x_1, \ldots, x_n)^T$ and $\tilde{x} = (\tilde{x}_1, \ldots, \tilde{x}_n)^T$ are the state vectors of the two systems, and $\{a_{ij}\}$ and $\{\tilde{a}_{ij}\}$ are the *parameters* of Eq. (2.31) and Eq. (2.32) respectively. Theorem 2.1 applied to Eqs. (2.31-2.32) results in

Corollary 2.2 *If $a_{ij} = \tilde{a}_{ij}$ for all i, j and*

$$\begin{aligned} \dot{x}_1 &= f_1(x, a_{11}, \ldots, a_{1m}, u, \tilde{u}, t) \\ &\vdots \\ \dot{x}_n &= f_n(x, a_{n1}, \ldots, a_{nm}, u, \tilde{u}, t) \end{aligned}$$

is asymptotically stable for all external inputs u and $\tilde{u}$, then the two systems (2.31-2.32) synchronize for arbitrary initial states $x(0)$ and $\tilde{x}(0)$.

A straightforward way to synchronize these two systems is the following two step approach. First, make sure that the two systems are synchronized when the parameters in the two systems are identical, i.e. the two systems are identical[§]. Second, adjust the parameters such that the corresponding parameters in the two systems approach each other.

Let us assume that the parameters in Eq. (2.31) are fixed. Assume that the two systems synchronize when the parameters are identical, i.e. the first step is satisfied. When the parameters in the two systems are different, the goal in the second step is to adapt the parameters $\{\tilde{a}_{ij}\}$ in Eq. (2.32) such that they approach the parameters $\{a_{ij}\}$ in Eq. (2.31). However, if we want to change the parameters of Eq. (2.32) without knowing any information about the parameters in Eq. (2.31), it is not clear that these parameters approach each other. If our goal is to synchronize Eqs. (2.31-2.32), what we want to do is to change $\{\tilde{a}_{ij}\}$ such that $\tilde{x}$ approaches x. For certain systems, it could be true that when $\tilde{x}$ approaches x, it is necessary that $\tilde{a}_{ij} \to a_{ij}$. However, in Section 2.7.4, we show an example where although the synchronization error are small, the parameters in the two systems are not matched (although they do oscillate near each other). This is an example of weak coupling where the two-step approach fails; the first step is not satisfied, and thus the second step cannot be used to achieve synchronization. Fortunately, this two-step approach is not always

---

[§]Thus when $a_{ij} = \tilde{a}_{ij}$, we have $\tilde{x} \to x$ as $t \to \infty$.

necessary for synchronization, and synchronization is still possible by the use of an adaptive controller.

Huberman and Lumer [47] proposed a class of adaptive algorithms for changing the parameters $\tilde{a}_{ij}$ in discrete time systems. For continuous time systems such as Eqs. (2.31-2.32), these algorithms take on the following form:

$$\dot{\tilde{a}}_{ij} = -\delta_{ij} r_i \text{sgn}\left(\frac{d\dot{r}_i}{d\tilde{a}_{ij}}\right) \qquad (2.33)$$

where $r_i = x_i - \tilde{x}_i$ and $\delta_{ij} > 0$ and $\frac{d\dot{r}_i}{d\tilde{a}_{ij}} = -\frac{d\dot{\tilde{x}}_i}{d\tilde{a}_{ij}}$ is defined as $-\frac{\partial f_i}{\partial \tilde{a}_{ij}}$. This algorithm was also used successfully in [49].

2.7.1 A general adaptive scheme

Definition 2.10 A function $\mu : R \to R$ is *strictly passive* [19] if it lies in the open first and third quadrants. Thus μ is strictly passive if and only if $\mu(0) = 0$ and $x\mu(x) > 0$ for all $x \neq 0$. A function $\mu : R \to R$ is in class $\mathcal{D}$ if it is strictly passive and there exists a strictly increasing function ψ such that $\psi(0) = 0$ and $|\mu(x)| \geq |\psi(x)|$ for all x.

An example of a function μ in class $\mathcal{D}$ is shown in Fig. 2.18. Many strictly passive functions belong to class $\mathcal{D}$. An example of a strictly passive function which is not in class $\mathcal{D}$ is a function which approaches the x-axis as $x \to \infty$ (Fig. 2.19).

Consider the general adaptive scheme given by:

$$\dot{\tilde{a}}_{ij} = -\delta_{ij} G\left(r_i, \frac{d\dot{r}_i}{d\tilde{a}_{ij}}\right) \qquad (2.34)$$

Note that Eq. (2.33) is in this form when $G\left(r_i, \frac{d\dot{r}_i}{d\tilde{a}_{ij}}\right) = r_i \text{sgn}\left(\frac{d\dot{r}_i}{d\tilde{a}_{ij}}\right)$. Let us choose the following function for G instead [53]:

$$G\left(r_i, \frac{d\dot{r}_i}{d\tilde{a}_{ij}}\right) = \mu(r_i)\left(\frac{d\dot{r}_i}{d\tilde{a}_{ij}}\right) = -\mu(r_i)\left(\frac{d\dot{\tilde{x}}_i}{d\tilde{a}_{ij}}\right) \qquad (2.35)$$

where μ is a function of class $\mathcal{D}$.

Some candidates for μ include the signum function $\text{sgn}(x)$, the identity function, $\text{sgn}(x)(e^{|x|} - 1)$, and x^n for n odd and positive. When μ is the

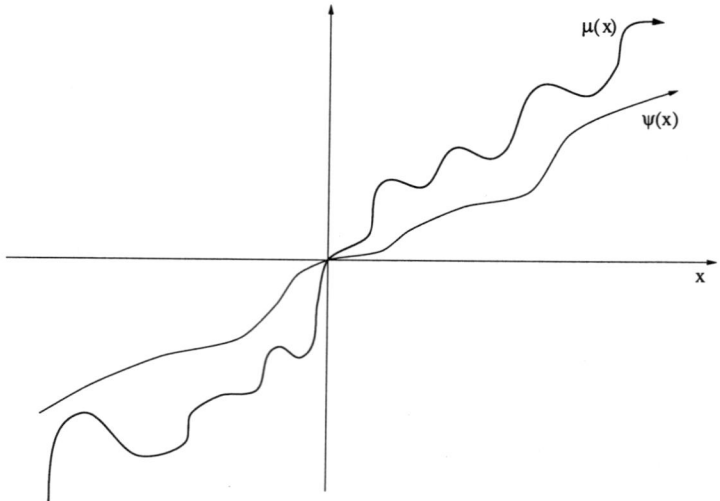

Fig. 2.18 Function μ belongs to class $\mathcal{D}$. Function ψ is a strictly increasing function such that $|\mu(x)| \geq |\psi(x)|$ for all x.

signum function, the resulting G becomes

$$G\left(r_i, \frac{d\dot{r}_i}{d\tilde{a}_{ij}}\right) = \text{sgn}(r_i)\left(\frac{d\dot{r}_i}{d\tilde{a}_{ij}}\right) \qquad (2.36)$$

There is an analogy between these schemes and the signed LMS schemes used in adaptive signal processing [54]. Eq. (2.33) corresponds to the Clipped LMS or Signed Regressor algorithm, while Eq. (2.36) corresponds to the Pilot LMS or Signed Error algorithm.

2.7.2 Two coupled nonlinear systems with linear parameters

One advantage of adaptation scheme (2.35) over (2.33) is that under certain conditions theoretical results can be derived for the class of nonlinear systems with linear parameters. Many chaotic systems in the literature, including those in Appendix D, belong to this class (see [55] for other examples of such chaotic systems).

We consider the class of systems where the parameters in the system are linear factors. The state equations for two such systems coupled together are given by:

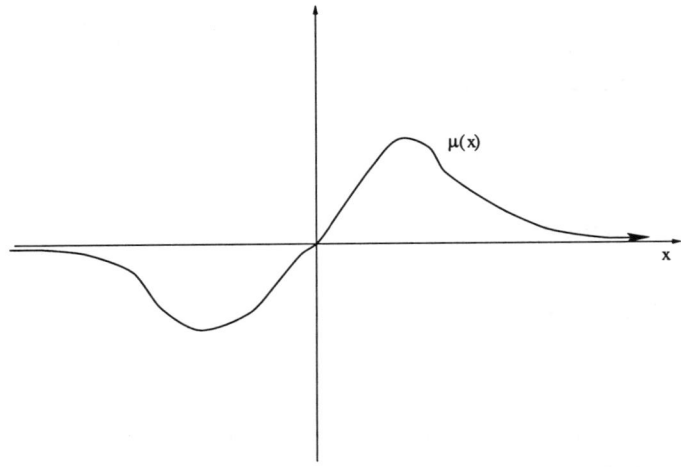

Fig. 2.19 The strictly passive function μ does not belong to class $\mathcal{D}$ since it approaches the x-axis as $x \to \pm\infty$.

$$\begin{aligned}
\dot{x}_1(t) &= d_1(x) + \sum_{j=1}^{m} a_{1j} f_{1j}(x) + u_1 + k_1(x) - k_1(\tilde{x}) \\
&\vdots \\
\dot{x}_n(t) &= d_n(x) + \sum_{j=1}^{m} a_{nj} f_{nj}(x) + u_n + k_n(x) - k_n(\tilde{x})
\end{aligned} \quad (2.37)$$

$$\begin{aligned}
\dot{\tilde{x}}_1(t) &= d_1(\tilde{x}) + \sum_{j=1}^{m} \tilde{a}_{1j} f_{1j}(\tilde{x}) + u_1 + h_1(\tilde{x}) - h_1(x) \\
&\vdots \\
\dot{\tilde{x}}_n(t) &= d_n(\tilde{x}) + \sum_{j=1}^{m} \tilde{a}_{nj} f_{nj}(\tilde{x}) + u_n + h_n(\tilde{x}) - h_n(x)
\end{aligned} \quad (2.38)$$

where $x = (x_1, \ldots, x_n)^T$ and $\tilde{x} = (\tilde{x}_1, \ldots, \tilde{x}_n)^T$ are the state vectors of Eq. (2.37) and Eq. (2.38) respectively. We assume that d_i, f_{ij} are nonlinear functions which are continuously differentiable. The vector $u = (u_1, \ldots, u_n)^T$ constitutes the external inputs to the two systems. Thus both systems receive the same external input. The scalars a_{ij} and $\tilde{a}_{ij}$ are the *parameters* of Eq. (2.37) and Eq. (2.38), respectively. The continuous functions h_i and k_i denote the additive coupling between the two systems. The functions d_i denote the part of the system which is identical in the two systems. There is no need to adapt this part of the system. For example, this part can contain parameters which are identical in both systems.

We assume that the parameters a_{ij} in Eq. (2.37) are constant, and try to adjust $\tilde{a}_{ij}$ adaptively such that $\tilde{x}$ approaches x as $t \to \infty$. Applying the adaptation rule (2.35) to this system results in:

$$\dot{\tilde{a}}_{ij} = \delta_{ij}\mu(x_i - \tilde{x}_i)f_{ij}(\tilde{x}) \tag{2.39}$$

for some constants $\delta_{ij} > 0$.

Definition 2.11 *A function $\mu : R \to R$ is odd-symmetric if $\mu(x) = -\mu(-x)$ for all x, i.e. the graph of μ is symmetric with respect to the origin.*

Lemma 2.1 *Let $\mu : R \to R$ be an odd-symmetric increasing function. Let g_i be a set of arbitrary real-valued continuously differentiable functions. If the functions g_i have bounded derivatives, then*

$$\left| \sum_i \mu(h_i) \left(g_i(x) - g_i(x+h) \right) \right| \leq p \sum_i \mu(h_i) h_i$$

for some $p > 0$ and for all x, h where $h = (h_1, \cdots, h_n)^T$.

Proof: By hypothesis, there exists $p > 0$ such that $|g_k(x) - g_k(x+h)| \leq p \max_i |h_i|$ for all k, x, h. Let j be the index such that $\max_i |h_i| = |h_j|$. Then

$$\begin{aligned}
|\mu(h_k)(g_k(x) - g_k(x+h))| &\leq |\mu(h_k)||g_k(x) - g_k(x+h)| \\
&\leq p|\mu(h_k)||h_j| \leq p|\mu(h_j)||h_j| = p\mu(h_j)h_j \\
&\leq p \sum_i \mu(h_i) h_i
\end{aligned}$$

Note that $|\mu(h_k)| \leq |\mu(h_j)|$ since μ is odd-symmetric and increasing. By summing over all indices k and redefining p we obtain the desired result. □

The following theorem gives theoretical justifications for the adaptive scheme in Eq. (2.35) by showing that the two systems synchronize when μ is a piecewise-continuous, odd-symmetric and increasing function in class $\mathcal{D}$ such that $x\mu(x)$ is continuous [53]:

Theorem 2.14 *Consider the adaptive controllers for $\tilde{a}_{ij}$ given by Eq. (2.39) where μ is a piecewise-continuous, odd-symmetric and increasing function in class $\mathcal{D}$ such that $x\mu(x)$ is continuous. We also assume that*

k_i's and h_i's are linear functions of the form $k_i(x) = k_i x_i$, $h_i(x) = h_i x_i$ and adapt the coefficients using the following controllers:

$$\dot{k}_i = -\kappa_i^k \mu(x_i - \tilde{x}_i)(x_i - \tilde{x}_i), \qquad \dot{h}_i = -\kappa_i^h \mu(x_i - \tilde{x}_i)(x_i - \tilde{x}_i) \qquad (2.40)$$

where $\kappa_i^k + \kappa_i^h > 0$.

If the Jacobian matrix of

$$\begin{pmatrix} d_1 + \sum_j a_{1j} f_{1j} \\ \vdots \\ d_n + \sum_j a_{nj} f_{nj} \end{pmatrix} \qquad (2.41)$$

as a function of x is bounded, then $x - \tilde{x} \to 0$ as $t \to \infty$.

Proof: Define $\Psi(x) = \int_0^x \mu(\tau) d\tau$. The function $x\Psi(x)$ is strictly increasing since μ belongs to class $\mathcal{D}$.

By Lemma 2.1

$$\left| \sum_i \mu(x_i - \tilde{x}_i) \left(\left(d_i + \sum_j a_{ij} f_{ij} \right)(x) - \left(d_i + \sum_j a_{ij} f_{ij} \right)(\tilde{x}) \right) \right|$$

is less than $p \sum_i \mu(x_i - \tilde{x}_i)(x_i - \tilde{x}_i)$ for some $p > 0$. Consider a set of positive constants $\{k_1^{\max}, \ldots, k_n^{\max}\}$ such that $k_i^{\max} > p$ for all i.

Construct the following Lyapunov function:

$$V = \sum_i \Psi(x_i - \tilde{x}_i) + \frac{1}{2} \sum_{ij} \frac{1}{\delta_{ij}} (a_{ij} - \tilde{a}_{ij})^2 + \frac{1}{2} \sum_i \frac{1}{\kappa_i^k + \kappa_i^h} (k_i^{\max} + k_i + h_i)^2$$

Since $x\Psi(x)$ is strictly increasing and $\Psi(0) = 0$, it follows that $\Psi(x) > 0$ for all $x \neq 0$. If we define $a(x) = \min(\Psi(x), \Psi(-x))$ and $b(x) = \max(\Psi(x), \Psi(-x))$, it follows that a and b are functions of class K (Appendix C) and $a(|x|) \leq \Psi(x) \leq b(|x|)$.

By differentiating V along the trajectories, we obtain

$$\begin{aligned}
\dot{V} &= \sum_i \mu(x_i - \tilde{x}_i)(\dot{x}_i - \dot{\tilde{x}}_i) + \sum_{ij} \frac{1}{\delta_{ij}} (a_{ij} - \tilde{a}_{ij})(-\dot{\tilde{a}}_{ij}) \\
&\quad - \sum_i (k_i^{\max} + k_i + h_i) \mu(x_i - \tilde{x}_i)(x_i - \tilde{x}_i) \\
&= \sum_i \mu(x_i - \tilde{x}_i)(d_i(x) - d_i(\tilde{x}) + (k_i + h_i)(x_i) - (k_i + h_i)(\tilde{x}_i)) \\
&\quad + \sum_{ij} \mu(x_i - \tilde{x}_i)(a_{ij} f_{ij}(x) - \tilde{a}_{ij} f_{ij}(\tilde{x})) \\
&\quad - \sum_{ij} (a_{ij} - \tilde{a}_{ij})(\mu(x_i - \tilde{x}_i) f_{ij}(\tilde{x})) \\
&\quad - \sum_i (k_i^{\max} + k_i + h_i) \mu(x_i - \tilde{x}_i)(x_i - \tilde{x}_i) \\
&= \sum_i \mu(x_i - \tilde{x}_i)(d_i(x) - d_i(\tilde{x}) - k_i^{\max}(x_i - \tilde{x}_i)) \\
&\quad + \sum_{ij} \mu(x_i - \tilde{x}_i)(a_{ij} f_{ij}(x) - a_{ij} f_{ij}(\tilde{x})) \\
&\leq \sum_i (p - k_i^{\max}) \mu(x_i - \tilde{x}_i)(x_i - \tilde{x}_i) \leq 0
\end{aligned}$$
(2.42)

Note that $(p - k_i^{\max}) < 0$ and $\mu(x_i - \tilde{x}_i)(x_i - \tilde{x}_i) = 0$ if and only if $x_i = \tilde{x}_i$. The values of $\tilde{a}_{ij}$ and $k_i + h_i$ are bounded since otherwise V will go to infinity. It can be shown that k_i and h_i are bounded [52] and the theorem then follows from Theorem C.2. □

Remark: It is clear from the proof of Theorem 2.14 that this result also holds if instead of adapting the coupling coefficients k_i and h_i we keep them fixed at values such that $k_i + h_i < -p$ for all i.

2.7.3 Two coupled nonlinear systems with multiplicative parameters

When the parameters are not linear, but multiplicative, a simple change of variables allow us to use Theorem 2.14 to design adaptive controllers for this case as well.

The state equations for two such systems coupled together are given by:

$$\begin{aligned}
\dot{x}_1(t) &= d_1(x) + \sum_{j=1}^m g_{1j}(a_{1j}) f_{1j}(x) + u_1 + k_1(x_1 - \tilde{x}_1) \\
&\vdots \\
\dot{x}_n(t) &= d_n(x) + \sum_{j=1}^m g_{nj}(a_{nj}) f_{nj}(x) + u_n + k_n(x_n - \tilde{x}_n)
\end{aligned}$$
(2.43)

$$\begin{aligned}
\dot{\tilde{x}}_1(t) &= d_1(\tilde{x}) + \sum_{j=1}^m g_{1j}(\tilde{a}_{1j}) f_{1j}(\tilde{x}) + u_1 + h_1(\tilde{x}_1 - x_1) \\
&\vdots \\
\dot{\tilde{x}}_n(t) &= d_n(\tilde{x}) + \sum_{j=1}^m g_{nj}(\tilde{a}_{nj}) f_{nj}(\tilde{x}) + u_n + h_n(\tilde{x}_n - x_n)
\end{aligned}$$
(2.44)

where $x = (x_1, \ldots, x_n)^T$ and $\tilde{x} = (\tilde{x}_1, \ldots, \tilde{x}_n)^T$ are the state vectors of

Eq. (2.43) and Eq. (2.44) respectively. The scalars a_{ij} and $\tilde{a}_{ij}$ are the parameters of the two systems, respectively. We assume that g_{ij}, d_i, f_{ij} are nonlinear functions which are continuously differentiable. As in the statement of Theorem 2.14 the coupling is linear and connects only corresponding state variables.

We assume that the parameters a_{ij} in Eq. (2.43) are constant, and try to adapt $\tilde{a}_{ij}$, k_i and h_i such that $\tilde{x}$ approaches x as $t \to \infty$.

Corollary 2.3 *Suppose that g'_{ij} is never zero¶. Consider the adaptive controllers for $\tilde{a}_{ij}$ given by*

$$\dot{\tilde{a}}_{ij} = \delta_{ij} \frac{\mu(x_i - \tilde{x}_i) f_{ij}(\tilde{x})}{g'_{ij}(\tilde{a}_{ij})} \qquad (2.45)$$

for some constants $\delta_{ij} > 0$ where μ is a piecewise-continuous, odd-symmetric and increasing function in class $\mathcal{D}$ such that $x\mu(x)$ is continuous. We adapt k_i and h_i using the adaptive controllers in Eq. (2.40).

If the Jacobian matrix of

$$\begin{pmatrix} d_1 + \sum_j g_{1j}(a_{1j}) f_{1j} \\ \vdots \\ d_n + \sum_j g_{nj}(a_{nj}) f_{nj} \end{pmatrix} \qquad (2.46)$$

as a function of x is bounded, then $x - \tilde{x} \to 0$ as $t \to \infty$.

Proof: Using the change of variables $\tilde{g}_{ij} = g_{ij}(\tilde{a}_{ij})$ and applying Theorem 2.14, we see that the following controllers for $\tilde{g}_{ij}$ will imply the conclusions of the corollary:

$$\dot{\tilde{g}}_{ij} = \delta_{ij} \mu(x_i - \tilde{x}_i) f_{ij}(\tilde{x}) \qquad (2.47)$$

Since $\dot{\tilde{g}}_{ij} = g'_{ij}(\tilde{a}_{ij}) \dot{\tilde{a}}_{ij}$, the corollary is proved. □

Some examples for g_{ij} include $g_{ij}(x) = \frac{1}{x}$ for $x > 0$ and $g_{ij}(x) = x$. For example, in the system $\dot{v} = -\frac{v^3}{R}$, the parameter R is multiplicative with $g(R) = \frac{1}{R}$.

Remark: if we apply Eq. (2.35) to system (2.43-2.44), we obtain the following adaptive controllers for $\tilde{a}_{ij}$:

$$\dot{\tilde{a}}_{ij} = \delta_{ij} \mu(x_i - \tilde{x}_i) g'_{ij}(\tilde{a}_{ij}) f_{ij}(\tilde{x})$$

which differs from Eq. (2.45) by a (time-varying) positive factor of $(g'_{ij}(\tilde{a}_{ij}))^2$.

¶This implies that g_{ij} is one-to-one.

2.7.4 Examples

In this section, we illustrate the results in the previous sections via computer simulations using Chua's oscillator. In particular, we choose several candidates for μ from the class $\mathcal{D}$ and use them to synchronize two nonidentical Chua's oscillators.

For simplicity, in contrast to Theorem 2.14 we will not use adaptive coupling. In particular, we will use a fixed coupling between only the x state variables in the two systems. In fact, we show that even when the coupling is not strong enough to synchronize the two systems when the parameters are identical, the use of adaptive controllers can still maintain the synchronization in some sense. In [51] it was shown that an adaptive controller for channel gain compensation also exhibits this phenomenon.

We will only adapt the parameter α and assume that all other parameters are matched. The dimensionless state equations used for synchronizing two Chua's oscillators are given by:

$$\begin{align}
\frac{dx}{dt} &= \alpha(y - x - f(x)) \\
\frac{dy}{dt} &= x - y + z \\
\frac{dz}{dt} &= -\beta y - \gamma z \\
\frac{d\tilde{x}}{dt} &= \tilde{\alpha}(\tilde{y} - \tilde{x} - f(\tilde{x})) + c(x - \tilde{x}) \\
\frac{d\tilde{y}}{dt} &= \tilde{x} - \tilde{y} + \tilde{z} \\
\frac{d\tilde{z}}{dt} &= -\beta\tilde{y} - \gamma\tilde{z} \\
\frac{d\tilde{\alpha}}{dt} &= \delta\mu(x - \tilde{x})(\tilde{y} - \tilde{x} - f(\tilde{x}))
\end{align} \tag{2.48}$$

and

$$f(x) = bx + \frac{1}{2}(a - b)(|x + 1| - |x - 1|)$$

The fixed parameters are chosen as $\alpha = 10$, $\beta = 15.6$, $\gamma = 0.001$, $a = -1.14$, $b = -0.714$.

In Fig. 2.20 we show $\tilde{\alpha}$ when $\delta = 100$, $\mu(x) = x$ and $c = -20a$. The parameter a corresponds to the slope of the middle segment of the piecewise linear function f. $\tilde{\alpha}$ is initially chosen to be 10^{-5}.

In Fig. 2.21 we show $\tilde{\alpha}$ when $c = -20a$, $\delta = 10$ and $\mu(x) = \text{sgn}(x)$. $\tilde{\alpha}$ is initially chosen to be 10^{-5}.

In Fig. 2.22 we show $\tilde{\alpha}$ when $c = -20a$, $\delta = 1$ and $\mu(x) = \text{sgn}(x)(e^{|x|} - 1)$. $\tilde{\alpha}$ is initially chosen to be 10^{-5}. We see that in all these cases $\tilde{\alpha}$ approaches $\alpha = 10$.

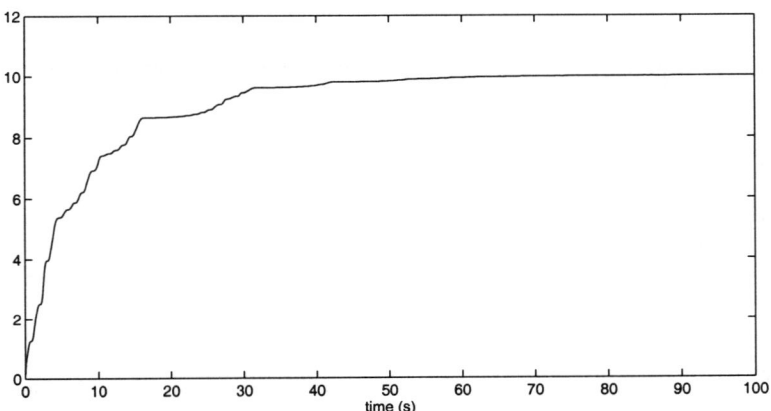

Fig. 2.20 $\tilde{\alpha}$ as a function of time, with $c = -20a$, $\delta = 100$ and $\mu(x) = x$.

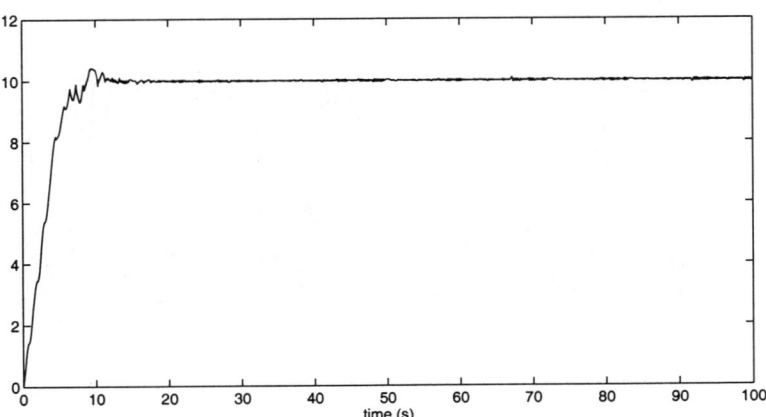

Fig. 2.21 $\tilde{\alpha}$ as a function of time, with $c = -20a$, $\delta = 10$ and $\mu(x) = \text{sgn}(x)$.

When we set $c = -2a$, the resulting coupling is not enough to synchronize the two systems even when $\alpha = \tilde{\alpha}$. In fact, simulations show that when $\alpha = \tilde{\alpha}$, $\tilde{x}$, $\tilde{y}$, and $\tilde{z}$ will diverge to infinity. However, when $\delta = 10$ and $\mu(x) = \text{sgn}(x)$, the adaptive controller for $\tilde{\alpha}$ will keep $\tilde{x}$ in phase with x. Figs. 2.23-2.25 compare the state variables in the two systems. We see that y and $\tilde{y}$ (and also z and $\tilde{z}$) are less similar to each other than x and $\tilde{x}$.

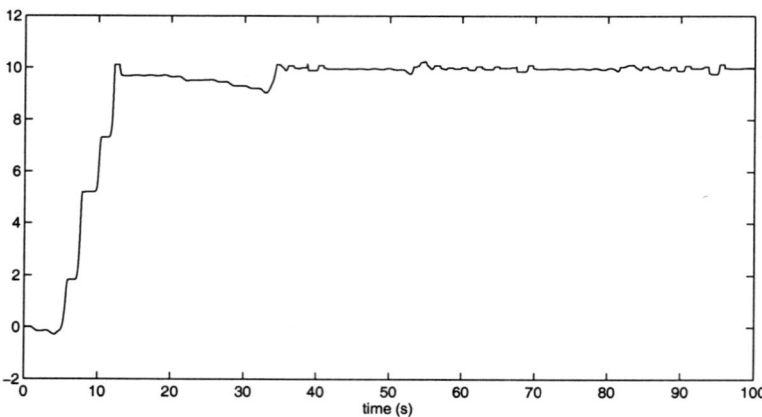

Fig. 2.22 $\tilde{\alpha}$ as a function of time, with $c = -20a$, $\delta = 1$ and $\mu(x) = \text{sgn}(x)(e^{|x|} - 1)$.

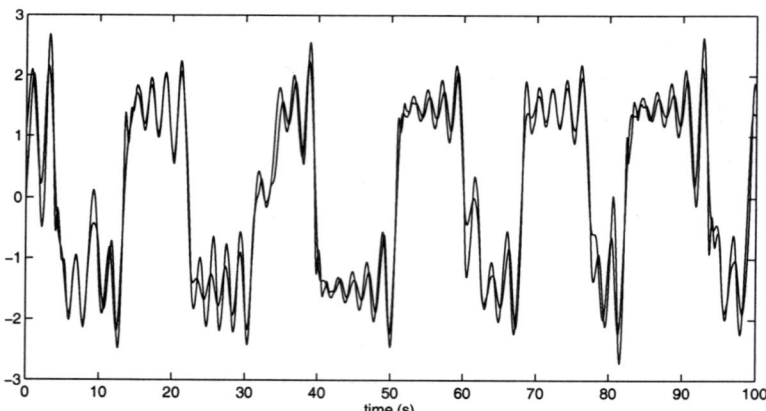

Fig. 2.23 x and $\tilde{x}$ as functions of time. $c = -2a$, $\delta = 10$, $\mu(x) = \text{sgn}(x)$.

In Fig. 2.26 we show $\tilde{\alpha}$ as a function of time. We see that $\tilde{\alpha}$ does not converge towards α, but fluctuates around a value below $\alpha = 10$. This is an example where $\tilde{\alpha}$ cannot converge towards α for the system to synchronize, since the coupling is too weak. Thus the time-varying mismatch in parameters is used here to *reduce* the synchronization error rather than to increase it.

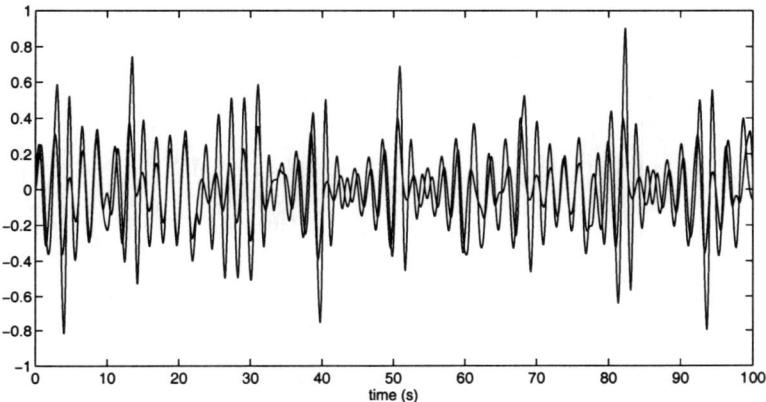

Fig. 2.24 y and $\tilde{y}$ as functions of time. $c = -2a$ $\delta = 10$, $\mu(x) = \text{sgn}(x)$.

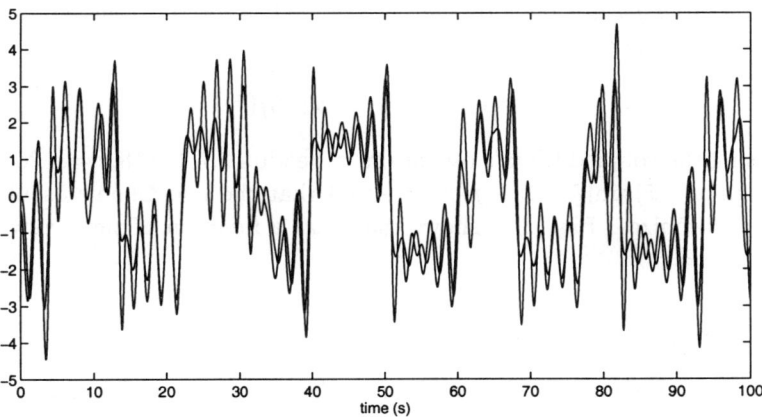

Fig. 2.25 z and $\tilde{z}$ as functions of time. $c = -2a$ $\delta = 10$, $\mu(x) = \text{sgn}(x)$.

2.7.5 *A generalization of the scheme in Eq. (2.35)*

A simple generalization of the scheme in Eq. (2.35) is to use the following G in Eq. (2.33):

$$G\left(r_i, \frac{d\dot{r}_i}{d\tilde{a}_{ij}}\right) = \mu_1(r_i)\mu_2\left(\frac{d\dot{r}_i}{d\tilde{a}_{ij}}\right) = -\mu_1(r_i)\mu_2\left(\frac{d\dot{\tilde{x}}_i}{d\tilde{a}_{ij}}\right)$$

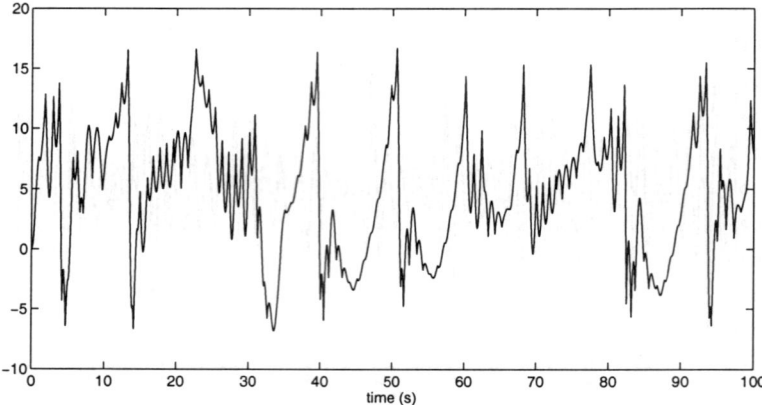

Fig. 2.26 $\tilde{\alpha}$ as a function of time. $c = -2a$ $\delta = 10$, $\mu(x) = \text{sgn}(x)$. We see that $\tilde{\alpha}$ fluctuates around a value below $\alpha = 10$.

where μ_1 and μ_2 are functions in class $\mathcal{D}$. Thus for the coupled systems (2.37)-(2.38) the controllers for $\tilde{a}_{ij}$ are:

$$\dot{\tilde{a}}_{ij} = \delta_{ij}\mu_1(x_i - \tilde{x}_i)\mu_2(f_{ij}(\tilde{x}))$$

Using the coupled Chua's oscillators in Section 2.7.4 with $c = -20a$ and $\dot{\tilde{\alpha}} = \text{sgn}(x - \tilde{x})\text{sgn}(\tilde{y} - \tilde{x} - f(\tilde{x}))$ we find that the two Chua's oscillators still synchronizes. Fig. 2.27 shows that $\tilde{\alpha}$ as a function of time converges towards $\alpha = 10$.

2.7.6 Adaptive observers

In Section 2.3 it was noted that observer design can be useful in designing synchronizing systems. Similarly, results in adaptive observer design can be useful in designing adaptive controllers for synchronizing coupled systems. In the adaptive synchronization scheme above, the entire state was needed in constructing the coupling and the adaptive controllers. When only a partial state is used, synchronized coupling is still possible for the case when the adaptive parameters appear as linear terms of nonlinear functions of this partial state. Consider the following result from adaptive observer design [56] when the system is in a canonical adaptive observer form where the unknown states are in linear terms and the adaptive parameters are linear with respect to the known states:

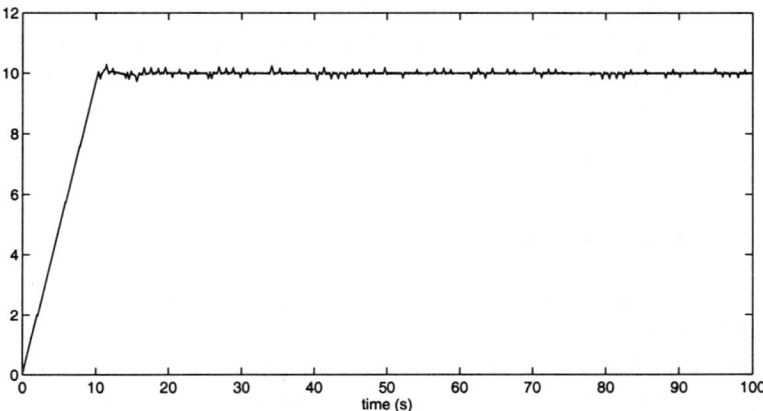

Fig. 2.27 $\tilde{\alpha}$ as a function of time with $c = -20a$ and $\dot{\tilde{\alpha}} = \text{sgn}(x - \tilde{x})\text{sgn}(\tilde{y} - \tilde{x} - f(\tilde{x}))$.

Theorem 2.15 *Consider the system*

$$\begin{aligned} \dot{x} &= Ax + f(y,t) + b\beta^T(t)a \\ y &= cx \end{aligned} \quad (2.49)$$

where $A = \begin{pmatrix} 0 & 1 & 0 & \ldots & 0 \\ 0 & 0 & 1 & \ldots & 0 \\ & & \ddots & & \\ 0 & 0 & 0 & \ldots & 1 \\ 0 & 0 & 0 & \ldots & 0 \end{pmatrix}$, $c = (1, 0, \ldots, 0)$, $b = (b_1, b_2, \ldots, b_n)^T$,

$a = (a_1, a_2, \ldots a_n)^T$ *is a vector of parameters and* $\beta(t)$ *is a vector common to both systems. For instance,* $\beta(t)$ *can be a function of the output* y. *If* $\beta(t)$ *is bounded,* $b_1 > 0$ *and the polynomial* $b_1 s^{n-1} + \ldots + b_n$ *has only roots in the open left half plane, then the following system is an adaptive observer in the sense that* $\tilde{x} - x \to 0$ *as* $t \to \infty$:

$$\begin{aligned} \dot{\tilde{x}} &= A\tilde{x} + f(y,t) + b\beta^T(t)\tilde{a} + k(y - c\tilde{x}) & (2.50) \\ \dot{\tilde{a}} &= \Gamma\beta(t)(y - c\tilde{x}) & (2.51) \end{aligned}$$

where Γ *is a positive definite symmetric matrix and the vector* k *is given by* $k = \frac{1}{b_1}(Ab + \lambda b)$ *for* $\lambda > 0$.

In [56] sufficient and necessary conditions are given under which a system with linear parameters can be transformed into the canonical adaptive observer form by a state transformation.

By encoding the information signal as time varying parameters in the transmitter system, adaptive synchronization can be used as a communication system if the estimated parameters track the parameters ($\tilde{a} \to a$) [57].

2.8 Discrete-time systems

So far, all the circuits and systems considered are continuous-time systems defined by ordinary differential equations. There are also many chaotic systems which are modelled as discrete-time systems of the form $x(k+1) = f(x(k), k)$, including the logistic map and the Hénon map [58]. An alternative notation is $x(k+1) = f^{(k)}(x(k))$. Many of the results in this chapter have a discrete-time counterpart. For instance, a discrete-time version of Theorem 2.1 is

Theorem 2.16 *The system*

$$\begin{aligned} x(k+1) &= f(x(k), x(k), y(k), k) \\ y(k+1) &= f(y(k), x(k), y(k), k) \end{aligned} \quad (2.52)$$

synchronizes in the sense that $\|x(k) - y(k)\| \to 0$ as $k \to \infty$, if $x(k+1) = f(x(k), u(k), v(k), k)$ is asymptotically stable for every $u(k)$ and $v(k)$.

Theorem 2.17 *If $f(x, k)$ satisfies:*

$$(f(x,k) - f(y,k))^T V (f(x,k) - f(y,k)) \leq \gamma (x-y)^T V (x-y)$$

for some positive definite matrix V, then $x(k+1) = f(x(k), k) + \eta(k)$ is asymptotically stable for all $\eta(k)$ if $0 \leq \gamma < 1$.

Proof: Let $y(k+1) = f(y(k), k) + \eta(k)$. It's clear that

$$(x(k+n) - y(k+n))^T V (x(k+n) - y(k+n)) \leq \gamma^n (x(k) - y(k))^T V (x(k) - y(k))$$

which vanishes as $n \to \infty$. By the positive definiteness of V this implies that $\|x(k+n) - y(k+n)\| \to 0$ as $n \to \infty$. □

Therefore, a sufficient condition for $x(k+1) = f(x(k), k)$ to be asymptotically stable is Lipschitz continuity of f with Lipschitz constant less than 1.

Theorems 2.11 and 2.12 are also valid for the discrete-time case. Instead of choosing a polynomial p with all roots in the open left half plane, a polynomial p is chosen with roots inside the open unit circle.

2.9 Further reading

The stability results discussed in this chapter are mainly chosen for their applicability to the coupled array case (Chapter 3). Nonlinear control and stability theory is an extensive field and the reader is referred to [21, 59, 60] for several comprehensive and modern treatments.

For further reading on other approaches to adaptive synchronization, the reader is referred to [61, 62]. The usefulness of observer design in synchronization is discussed in [12]. Some recent papers on synchronization of chaos can be found in [63, 64]. For recent progress on communication systems using chaos, see [32].

Chapter 3

Synchronization in Coupled Arrays of Chaotic Systems

In this chapter we extend some of the results for two coupled systems in Chapter 2 to arrays of coupled systems.

We assume a coupled array of identical systems where the state equations of each system are of the form

$$\dot{x}_i = \hat{f}(x_i, t) \tag{3.1}$$

and x_i is the state vector of the i-th system.

The array of coupled systems can be written in the following form:

$$\begin{aligned} \dot{x}_1 &= f_1(x_1, x_1, x_2, \ldots, x_n, t) \\ \dot{x}_2 &= f_2(x_2, x_1, x_2, \ldots, x_n, t) \\ &\vdots \\ \dot{x}_n &= f_n(x_n, x_1, x_2, \ldots, x_n, t) \end{aligned} \tag{3.2}$$

where f_i is obtained from $\hat{f}$ by adding the coupling between the systems. The first argument of f_i is the state x_i of the i-th system, and the other arguments are the state coupling from all the systems in the array, and the time dependence t.

Definition 3.1 A set of coupled chaotic circuits with state equations (3.2) *synchronizes* if the states of any two circuits, x_i and x_j, satisfy $\|x_i - x_j\| \to 0$ as $t \to \infty$.

For this definition to make sense, the states of the circuits must be of the same dimension which is true in our case since the systems are identical. The underlying connection topology of such coupled circuits can often be expressed as a connectivity graph or a connectivity hypergraph, and properties of these graphs will be useful in determining synchronization. The reader is referred to Appendix B for a summary of basic terminology of graph theory.

Definition 3.2 The *connectivity graph* of coupled array (3.2) is defined as the directed graph with vertex set $V = \{v_i\}$ and edge set E defined by

$$(v_i, v_j) \in E \Leftrightarrow v_i \neq v_j \text{ and } f_j \text{ depends on } x_i$$

For instance, the connectivity graph of the array in Fig. 3.1 is shown in Fig. 3.2 (drawn as an undirected graph).

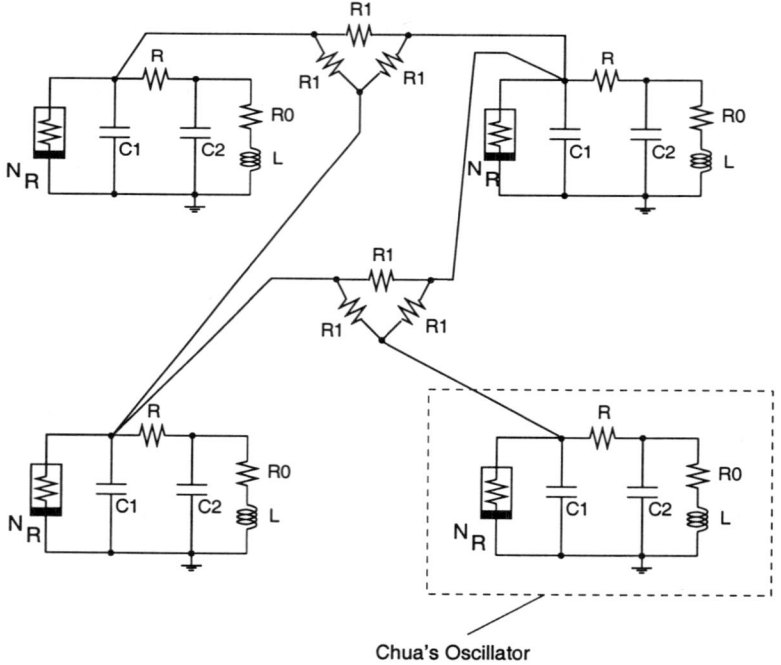

Fig. 3.1 Array of Chua's oscillators with uniform linear static coupling.

As in Chapter 2, the following theorem follows trivially from Definition C.1:

Theorem 3.1 *The coupled system (3.2) synchronizes if* $f = f_1 = f_2 =$

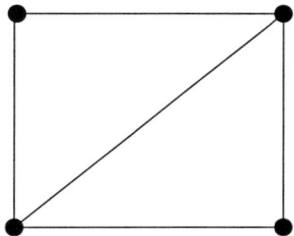

Fig. 3.2 Connectivity graph of the coupled array shown in Fig. 3.1.

$\ldots = f_n$ and $\dot{x} = f(x, \sigma_1(t), \ldots, \sigma_n(t), t)$ is asymptotically stable for all σ_i's.

Analogous to Chapter 2, Theorem 3.1 suggests the following way to design synchronizing arrays of chaotic circuits. Given a chaotic system $\dot{x} = \hat{f}(x, t)$, if $\hat{f}$ can be rewritten as

$$\hat{f}(x, t) = f(x, x, \ldots, x, t) \tag{3.3}$$

such that

$$\dot{x} = f(x, \sigma_1(t), \ldots, \sigma_n(t), t) \text{ is asymptotically stable for all } \sigma_i\text{'s} \tag{3.4}$$

then the coupled array of n systems

$$\begin{aligned} \dot{x}_1 &= f(x_1, x_1, x_2, \ldots, x_n, t) \\ &\vdots \\ \dot{x}_n &= f(x_n, x_1, x_2, \ldots, x_n, t) \end{aligned} \tag{3.5}$$

synchronizes. Furthermore, the consistency relation in Eq. (3.3) implies that at the synchronized state when $x_1(t) = x_2(t) = \ldots = x_n(t)$, the state vector $x_i(t)$ of each circuit follows the trajectory of the uncoupled system $\dot{x} = \hat{f}(x, t)$.

As in Chapter 2 we show here how $\hat{f}$ can be rewritten as f when there exists a stabilizing feedback term $g(x, t)$ such that $\dot{x} = \hat{f}(x, t) + g(x, t) + \eta(t)$ is asymptotically stable for all $\eta(t)$. Then f defined as

$$f(x, x_1, \ldots, x_n, t) = \hat{f}(x, t) + g(x, t) - \sum_{i=1}^{i=n} a_i g(x_i, t) + \eta(t) \tag{3.6}$$

satisfies the conditions in Eq. (3.4). The consistency relation (3.3) is satisfied when $\sum_{i=1}^{i=n} a_i = 1$ and $\eta(t) = 0$. Such a feedback term $g(x, t)$ can be found using the methods in Section 2.3.

For example, consider the fully coupled system of Chua's oscillators in Fig. 3.3. This can be written in the form of Eq. (3.2) if we choose

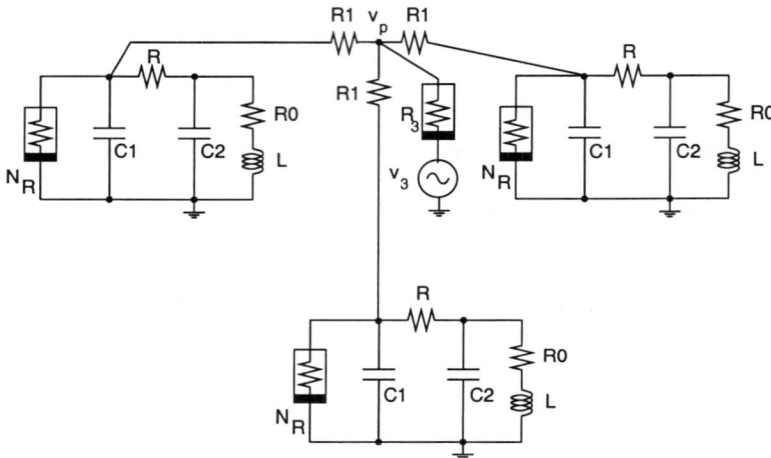

Fig. 3.3 An array of fully coupled Chua's oscillators where each Chua's oscillator is coupled to every other Chua's oscillator.

$$\hat{f}_i(x) = \begin{cases} \frac{1}{RC_1}((x^2 - x^1) - Rf(x^1)) \\ \frac{1}{RC_2}(x^1 - x^2 + Rx^3) \\ \frac{-1}{L}(x^2 + R_0 x^3) \end{cases} \tag{3.7}$$

where $x = (x^1, x^2, x^3)^T$ and

$$f_i(x, v_1, v_2, v_3, t) = \hat{f}_i(x) - \frac{1}{RC_1}(x_1, 0, 0)^T + \frac{1}{RC_1}(v_p(t), 0, 0)^T$$

In terms of Eq. (3.6), $g(x,t) = -\frac{1}{RC_1}(x_1,0,0)^T$, $a_i = 0$, and $\eta(t) = \frac{1}{RC_1}(v_p(t),0,0)^T$. For circuit parameters such that R, R_0, R_1, C_1, C_2, L are positive and the nonlinear resistor N_R has a globally Lipschitz characteristic, it can be shown from Section 2.3.3 that a small enough R_1 makes $\dot{x} = \hat{f}_1(x,t) + g(x,t) + \eta(t)$ asymptotically stable and thus the array synchronizes.

Note that in this example the consistency relation (3.3) is not satisfied and thus the synchronized systems do not follow the trajectories of an uncoupled system in general.

A subset of the systems with state equation (3.2) are the systems with static coupling, where the coupling elements do not exhibit dynamics of their own, e.g. when the circuits are coupled via memoryless devices. We describe the case of dynamic coupling (of which incidentally Fig. 3.3 is an example) in Chapter 4. Fig. 3.1 is an example of a statically coupled array of Chua's oscillators. The coupling elements consist of two-terminal linear resistors R_1.

3.1 Uniform linear static coupling

In this section we will study the class of statically coupled circuits where all the circuits are identical and the coupling is linear and uniform.

Definition 3.3 The Kronecker product or tensor product of matrices A and B is defined as:

$$A \otimes B = \begin{pmatrix} A_{11}B & \cdots & A_{1m}B \\ \vdots & \ddots & \\ A_{n1}B & & A_{nm}B \end{pmatrix}$$

where A is an n by m matrix, B is a p by q matrix and $A \otimes B$ is an np by mq matrix.

Lemma 3.1 Let p_1 and p_2 be two polynomials. If A, B, C are real symmetric matrices with eigenvectors a_i, b_i and c_i and corresponding eigenvalues λ_i^a, λ_i^b and λ_i^c respectively, then the symmetric matrix $p_1(A) \otimes B + p_2(A) \otimes C$ has a full set of eigenvectors $w_{ij} = a_i \otimes v_{ij}$ with corresponding eigenvalues λ_{ij} where v_{ij} are the eigenvectors of $p_1(\lambda_i^a)B + p_2(\lambda_i^a)C$ with corresponding eigenvalues λ_{ij}.

Proof: It's clear that $\{w_{ij}\}$ is a full set of eigenvectors.

$$\begin{aligned} (p_1(A) \otimes B + p_2(A) \otimes C) w_{ij} &= p_1(\lambda_i^a) a_i \otimes B v_{ij} + p_2(\lambda_i^a) a_i \otimes C v_{ij} \\ &= a_i \otimes (p_1(\lambda_i^a) B v_{ij} + p_2(\lambda_i^a) C v_{ij}) \\ &= a_i \otimes \lambda_{ij} v_{ij} = \lambda_{ij} w_{ij} \end{aligned}$$

□

Definition 3.4 $A \otimes f(x_i, t)$ is defined as:

$$A \otimes f(x_i, t) = \begin{pmatrix} A_{11} f(x_1, t) + A_{12} f(x_2, t) + \cdots + A_{1m} f(x_m, t) \\ \vdots \\ A_{n1} f(x_1, t) + A_{n2} f(x_2, t) + \cdots + A_{nm} f(x_m, t) \end{pmatrix}$$

where A is an n by m matrix, f is a function into R^p and $A \otimes f(x_i, t)$ is a vector of length np. It can be shown that $(A \otimes B)(C \otimes f(x_i, t)) = (AC \otimes Bf(x_i, t))$. When $x = (x_1, x_2, \ldots, x_m)^T$ and f does not depend on t, we write $A \otimes f(x_i)$ as $(A \otimes f)x$.

Note that when f is linear operator in the form of a matrix B and $x = (x_1, x_2, \ldots, x_m)^T$ then $A \otimes f(x_i) = (A \otimes f)x = (A \otimes B)x$.

A general form for the state equations of a uniform linear statically coupled array of m identical systems is:

$$\dot{x} = \begin{pmatrix} \hat{f}(x_1, t) \\ \vdots \\ \hat{f}(x_m, t) \end{pmatrix} + (G \otimes D)x = I_m \otimes \hat{f}(x_i, t) + (G \otimes D)x \quad (3.8)$$

where $x = (x_1, \ldots, x_m)^T$.

The matrix G can be thought of as defining the coupling topology of the array while the matrix D defines the coupling between two circuits in the array. We call this type of coupling *uniform* since the matrix D is the same between any two coupled circuits. The state equations of Fig. 3.1 can be written in the form of Eq. (3.8) if f is given by $\hat{f}$ in Eq. (3.7), $D = (\frac{1}{R_1 C_1}, 0, 0)^T$ and

$$G = \begin{pmatrix} -2 & 1 & 1 & 0 \\ 1 & -4 & 2 & 1 \\ 1 & 2 & -4 & 1 \\ 0 & 1 & 1 & -2 \end{pmatrix}$$

Note that the connectivity graph of Eq. (3.8) is given by the graph of G (excluding the self-loops)*.

Definition 3.5 The set W consists of all zero row sum matrices which have only nonpositive off-diagonal elements.

---
*A graph of a square matrix A is given by (V, E) where $(v_i, v_j) \in E \Leftrightarrow A_{ij} \neq 0$.

Definition 3.6 The set W_i consists of all irreducible matrices in W.

For the coupled system (3.8) the following theorem gives sufficient conditions for synchronization [13, 65]:

Theorem 3.2 *Let T be a matrix such that $f(x,t) + Tx$ is V-uniformly decreasing for some symmetric positive definite matrix V. System (3.8) synchronizes in the sense that $x_i \to x_j$ as $t \to \infty$ if there exists a symmetric matrix U in W_i such that the matrix*

$$(U \otimes V)(G \otimes D - I \otimes T) \quad \text{is negative semidefinite.} \tag{3.9}$$

The matrix T can be thought of as a stabilizing linear state feedback matrix such that $\dot{x} = f(x,t) + Tx$ is asymptotically stable.

Proof: Construct the Lyapunov function $g(x) = \frac{1}{2}x^T(U \otimes V)x$. By Lemma A.12 U is positive semidefinite and has a zero eigenvalue of multiplicity 1 with eigenvector $(1,\ldots,1)^T$. Therefore g is zero if $x_i = x_j$ for all i,j and positive elsewhere. The derivative of g along trajectories of (3.8) is:

$$\dot{g} = x^T(U \otimes V)\dot{x} = x^T(U \otimes V)\begin{pmatrix} f(x_1,t) + Tx_1 \\ \vdots \\ f(x_m,t) + Tx_m \end{pmatrix}$$
$$+ x^T(U \otimes V)(G \otimes D - I \otimes T)x \tag{3.10}$$

By Lemma A.12 U can be decomposed as $M^T M$ where M is a matrix such that each row of M contains zeros and one entry α and one entry $-\alpha$ for some $\alpha \neq 0$. In other words, the first term of Eq. (3.10) is of the form $\sum \alpha_{ij}^2 (x_i - y_j)^T V(f(x_i,t) + Tx_i - f(x_j,t) - Tx_j)$ which is nonpositive by the V-uniformly decreasing property of $f+T$. In fact, by the irreducibility of U, this term is zero exactly when $x_i = x_j$ for all i,j. Since the second term of Eq. (3.10) is nonpositive by hypothesis, the theorem is proved by Lyapunov's direct method (Theorem C.4). □

Condition (3.9) depends on the properties of both G and D. By restricting to special cases for the matrix G, Theorem 3.2 can be simplified. In particular, for normal matrices G, the dependence on G can be separated from the dependence on D and a synchronization criterion can be obtained which depends on the matrix properties of G and of D separately [65]. Since G describes the coupling topology, this in effect decomposes the synchronization condition into a component which depends on G (the coupling topology) and a component which depends on D (the coupling matrix between two systems). This is useful in studying how the synchronization properties change by changing only the coupling topology or only the coupling between two circuits.

Corollary 3.1 Let $f(x,t)$ be Lipschitz continuous in x. If G is a symmetric matrix in W_i and D is symmetric negative definite, then there exists a constant $\alpha > 0$ such that the system (3.8) synchronizes if the nonzero eigenvalues of G are larger than α.

Proof: By Theorem 2.3, there exists $\alpha > 0$ such that $\dot{x} = f(x,t) + \alpha Dx$ is uniformly decreasing. Let $U = G$, $V = I$ and $T = \alpha D$. The matrix $(U \otimes V)(G \otimes D - I \otimes T)$ is equal to $G(G - \alpha I) \otimes D$ which is negative semidefinite if $G(G - \alpha I)$ is positive semidefinite. By the spectral mapping theorem it is clear that the eigenvalues of $G(G - \alpha I)$ are nonnegative if the nonzero eigenvalues of G are larger than or equal to α. □

In Section 3.3 we will see that this result is true for all Hurwitz and diagonalizable D.

3.1.1 G is normal

Suppose that G is a real normal matrix (i.e. $GG^T = G^T G$) such that $G + G^T$ has zero row sums, has nonpositive off-diagonal elements and is irreducible. We also assume that VD is a symmetric matrix. By choosing $U = (G + G^T)$, Theorem 3.2 is reduced to:

Corollary 3.2 Let T be a matrix such that $f(x,t) + Tx$ is V-uniformly decreasing for some symmetric positive V. Suppose G is a normal matrix such that $G + G^T \in W_i$. If D is a matrix such that VD is symmetric then system (3.8) synchronizes if for all eigenvalues λ_i of G not on the imaginary axis, the eigenvalues of $2Re(\lambda_i)VD - (VT + T^T V)$ are nonpositive.

Proof: By choosing $U = (G + G^T)$, the array synchronizes by Theorem 3.2 if $A = (G + G^T)G \otimes VD - (G + G^T) \otimes VT$ is negative semidefinite. By normality of G,

$$A + A^T = (G + G^T)^2 \otimes VD - (G + G^T) \otimes (VT + T^T V)$$

By Lemma 3.1 the eigenvalues of $A + A^T$ are the eigenvalues of $\mu_i^2 VD - \mu_i(VT + T^T V) = \mu_i(\mu_i VD - (VT + T^T V))$ where μ_i are the eigenvalues of $(G + G^T)$. Since $\mu_i \geq 0$, A is negative semidefinite if the eigenvalues of

$\mu_i VD - (VT + T^T V)$ are nonpositive for all $\mu_i > 0$. Since the eigenvalues of $G + G^T$ are of the form $2Re(\lambda_i)$ where λ_i are the eigenvalues of G [66], the proof is complete. □

Definition 3.7 S is the set of complex numbers λ such that all eigenvalues of $\lambda(VD + D^T V) - (VT + T^T V)$ are nonpositive.

Thus the array synchronizes if the real parts of the spectrum of G are in $S \cup \{0\}$. The requirement that G is normal is important for Corollary 3.2. If G is not normal then $G + G^T$ having constant row sums does not imply that G has constant row sums. Similarly $G + G^T$ being irreducible does not imply that G is irreducible. For example, consider

$$G = \begin{pmatrix} 1 & 0 & 0 \\ -2 & 2 & -2 \\ 0 & 0 & 1 \end{pmatrix} \quad (3.11)$$

This matrix G is not normal, does not have constant row sums and is not irreducible. Yet $G + G^T$ has zero row sums and is irreducible. It is easy to create examples where G not having constant row sums or G not being irreducible makes synchronization impossible. If G does not have constant row sums, this implies that the coupling terms added to each system is different and for many systems, $(x_i, x_i, \ldots, x_i)^T$ is therefore not a solution of Eq. (3.8) which means that synchronization is not possible. The graph of the reducible matrix G in Eq. (3.11) is shown in Fig. 3.4. It is clear that since system 1 and system 3 do not influence each other, they cannot synchronize, especially for chaotic systems exhibiting sensitive dependence on initial conditions, where coupling between two systems are needed for synchronization.

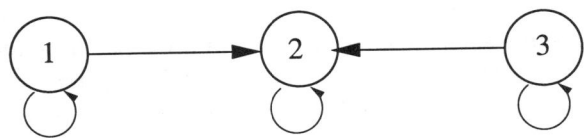

Fig. 3.4 Graph corresponding to coupling matrix G in Eq. (3.11).

But the normality of G ensures that the zero row sum property and the irreducibility property of $G + G^T$ persist in G (Lemma A.2). It is left as an exercise for the reader to determine how Corollary 3.2 can be used to prove some portion of Lemma A.2.

3.1.2 G is symmetric

Suppose G is a real irreducible symmetric matrix with zero row sums and nonpositive off-diagonal elements. Examples of such matrices include Laplacian matrices of connected graphs. By choosing $U = G$, Theorem 3.2 is reduced to the following:

Corollary 3.3 *Let T be a matrix such that $f(x,t) + Tx$ is V-uniformly decreasing for some symmetric positive definite V. Let G be a symmetric matrix in W_i. System (3.8) synchronizes if for all nonzero eigenvalues λ_i of G, the eigenvalues of $\lambda_i(VD + D^TV) - (VT + T^TV)$ are nonpositive.*

Proof: By choosing $U = G$, the array synchronizes by Theorem 3.2 if $A = G^2 \otimes VD - G \otimes VT$ is negative semidefinite.

$$A + A^T = G^2 \otimes (VD + D^TV) - G \otimes (VT + T^TV)$$

By Lemma 3.1 the eigenvalues of $A + A^T$ are the eigenvalues of $\lambda_i^2(VD + D^TV) - \lambda_i(VT + T^TV)$ where λ_i are the eigenvalues of G. Since $\lambda_i \geq 0$, A is negative semidefinite if the eigenvalues of $\lambda_i(VD + D^TV) - (VT + T^TV)$ are nonpositive for all $\lambda_i > 0$. □

In this case, the array synchronizes if the spectrum of G lies in $S \cup \{0\}$. Consider the special case where $D = \beta T$ for some $\beta > 0$, i.e. the coupling D is proportional to the linear stabilizing state feedback T. If VT is negative semidefinite, then by Lemma A.6 any matrix G satisfying condition (3.9) must necessarily be a matrix with constant row sums. Furthermore, the set S contains all positive numbers larger than or equal to $\frac{1}{\beta}$. In other words, the array synchronizes if all the nonzero eigenvalues of G are positive enough. In particular, Corollary 3.3 reduces to:

Corollary 3.4 *Let D be a matrix such that $f(x,t) + Dx$ is V-uniformly decreasing for some symmetric positive definite V. Let G be an symmetric matrix in W_i. If VD is negative semidefinite, then system (3.8) synchronizes if all nonzero eigenvalues of G are larger than or equal to 1.*

Similarly Corollary 3.2 reduces to:

Corollary 3.5 *Let D be a matrix such that $f(x,t) + Dx$ is V-uniformly decreasing for some symmetric positive definite V. Let G be a normal matrix such that $G + G^T \in W_i$. If VD is symmetric negative semidefinite, then system (3.8) synchronizes if all eigenvalues of G which are not imaginary have real parts larger than or equal to 1.*

It is easy to see that if VD is negative semidefinite and $f + D$ is V-uniformly decreasing, then $f+\mu D$ is V-uniformly decreasing for $\mu \geq 1$. This implies that increasing the coupling term preserves the synchronization in this type of coupling. Examples of cases where both V and VD are symmetric are when V and D are both diagonal and when D is a multiple of the identity matrix.

In Fig. 3.1, the matrix G is proportional to the Laplacian matrix of the underlying coupling graph. In this case Eq. (3.8) can be interpreted as follows: the individual systems $\dot{x}_i = \hat{f}(x_i, t)$ are the vertices of the graph and a linear coupling term $\alpha D(x_j - x_i)$ is added to the right hand side of $\dot{x}_i = \hat{f}(x_i, t)$ if and only if x_i and x_j are adjacent vertices. The smallest nonzero eigenvalue of the Laplacian matrix is the algebraic connectivity of the graph and gives a measure of how tightly coupled the graph is. Thus Corollary 3.4 implies that given a fixed coupling matrix D, the synchronization condition is proportional to the algebraic connectivity of the underlying coupling graph; the array synchronizes if the algebraic connectivity is large enough. We explore the relationship between synchronization and graph topology further in Chapter 5.

3.1.3 General G

When G is not normal, it is more difficult to determine what the best choice for U would be.

Definition 3.8 For a zero row sum matrix G, let $L(G)$ denote the set of eigenvalues of G which do not correspond to the eigenvector $(1, \ldots, 1)^T$.

In particular, if G has a zero eigenvalue of multiplicity 1, then $0 \notin L(G)$.

Definition 3.9 Let $\mu(G)$ be the supremum of all real numbers μ such that $U(G - \mu I)$ is positive semidefinite for some symmetric matrix U in W_i.

Note that if $U(G - \mu I) \geq 0$ for some $U \geq 0$, then $U(G - \lambda I) \geq 0$ for all $\lambda \leq \mu$. Furthermore, by Lemma A.6, $\mu(G)$ is only defined if G has constant row sums.

Corollary 3.6 *System (3.8) synchronizes if there exists some symmetric positive definite matrix V such that $f(x,t) + \mu(G)Dx$ is V-uniformly decreasing and VD is symmetric negative semidefinite.*

Proof: If $U(G - \mu(G)I) \geq 0$, then $U(G - \mu(G)I) \otimes VD \leq 0$ since $VD \leq 0$ is symmetric and the proof follows directly from Theorem 3.2. Otherwise, for small enough $\epsilon > 0$, $U(G - (\mu(G) - \epsilon)I) \geq 0$ and $f(x,t) + (\mu(G) - \epsilon)Dx$ is still V-uniformly decreasing and the proof follows from Theorem 3.2. $\square$

Thus for a fixed D, the larger $\mu(G)$ is, the easier it is to synchronize the array. Let us consider the problem of computing $\mu(G)$. The following theorem gives an upper bound on $\mu(G)$ when G is a zero row sum matrix.

Theorem 3.3 *Let G be a zero row sum matrix. If λ is a real eigenvalue in $L(G)$, then $\mu(G) \leq \lambda$.*

Proof: Let λ be a real eigenvalue of G in $L(G)$ with eigenvector v. Since v is not of the form $\gamma(1,1,\ldots,1)^T$, v is also not in the kernel of U. $(G - \mu I)v = (\lambda - \mu)v$, and thus $v^T U(G - \mu I)v = (\lambda - \mu)v^T Uv$. $v^T Uv > 0$ since v is not in the kernel of U. This in combination of the definition of $\mu(G)$ implies that $\lambda - \mu(G) \geq 0$. $\square$

We now show two classes of matrices with real eigenvalues where this upper bound is also a lower bound, i.e., $\mu(G)$ is equal to the least real eigenvalue in $L(G)$. The first class is the class of triangular zero row sum matrices.

Theorem 3.4 *If G is a triangular zero row sum matrix, then $\mu(G)$ is the least eigenvalue of $L(G)$.*

For instance, if G is an upper triangular zero row sum matrix, then $\mu(G)$ is equal to the least diagonal element of G, excluding the lower-right diagonal element.

Proof: Without loss of generality, suppose that G is an $n \times n$ upper triangular zero sum matrix:

$$G = \begin{pmatrix} a_{1,1} & a_{1,2} & \cdots & & & a_{1,n} \\ 0 & a_{2,2} & a_{2,3} & \cdots & & a_{2,n} \\ 0 & 0 & \ddots & & & \\ 0 & 0 & 0 & & a_{n-1,n-1} & -a_{n-1,n-1} \\ 0 & 0 & \cdots & & & 0 \end{pmatrix} \quad (3.12)$$

From Theorem 3.3, $\mu(G) \leq \min_{1 \leq i \leq n-1} a_{i,i}$. By Lemma A.3,

$$B = \begin{pmatrix} a_{1,1} & b_{1,2} & \cdots & \\ & a_{2,2} & b_{2,3} & \cdots \\ & & \ddots & \\ & & & a_{n-1,n-1} \end{pmatrix}$$

is an $(n-1) \times (n-1)$ upper triangular matrix which satisfies $CG = BC$ where C is as in Eq. (A.1). Let $\Delta = \mathrm{diag}(\alpha_1, \ldots \alpha_{n-1})$ where $\alpha_i > 0$ and $H = \Delta B \Delta^{-1}$. The (i,j)-th element of H is $b_{i,j} \frac{\alpha_i}{\alpha_j}$ if $j > i$ and 0 if $j < i$. Therefore, for each $\epsilon > 0$, if we choose α_j much larger than α_i for all $j > i$, then we can ensure that the (i,j)-th element of H has absolute value less than $\frac{2\epsilon}{n-2}$ for $j > i$. By Gershgorin's circle criterion the eigenvalues of $\frac{1}{2}(H + H^T)$ is larger than $\min_{1 \leq i \leq n-1} a_{i,i} - \epsilon$. Consider $U = C^T \Delta^2 C$. Since $\Delta C \in M_2$, $U \in W_i$ by Lemma A.8. Using Lemma A.11

$$U(G - \mu I) = C^T \Delta^2 C(G - \mu I) = C^T \Delta (H - \mu I) \Delta C$$

which is positive semidefinite if $H - \mu I$ is positive semidefinite. From the discussion above $H - \mu I \geq 0$ if $\mu \leq \min_{1 \leq i \leq n-1} a_{i,i} - \epsilon$. Therefore $\mu(G) \geq \min_{1 \leq i \leq n-1} a_{i,i}$. □

The second class of matrices where we can explicitly determine $\mu(G)$ is the class of symmetric matrices in W_i.

Theorem 3.5 *If $G \in W_i$ is symmetric, then $\mu(G)$ is the least nonzero eigenvalue of G.*

Proof: Let α be the least nonzero eigenvalue of G. If we choose $U = G$, then $U(G - \alpha I) = U^2 - \alpha U$ is a symmetric matrix whose eigenvalues are of the form $\lambda(\lambda - \alpha)$ for λ an eigenvalue of U. Since $\lambda \geq \alpha$ for $\lambda \neq 0$, this implies that $U(G - \alpha I)$ has only nonnegative eigenvalues and is thus positive semidefinite. This implies that $\alpha \leq \mu(G)$. Combine this Theorem 3.3 we get $\alpha = \mu(G)$. □

In general, the value of $\mu(G)$ can be computed using optimization techniques. The use of nonlinear optimization to find good matrices for constructing Lyapunov functions is a common technique in control system design and analysis [67]. In our case, $\mu(G)$ can be computed as

$$-\mu(G) = \min_{U = U^T \in W_i, U(G + \beta I) \geq 0} \beta$$

We conjecture that the bound in Theorem 3.3 also give the value of $\mu(G)$ for a class of nonsymmetric matrices G in W_i.

Conjecture 3.1 *Let G be a matrix in W_i. If there exists a real eigenvalue λ of G such that the real parts of all nonzero eigenvalues of G are larger than or equal to λ, then $\mu(G) = \lambda$.*

This conjecture was verified by numerical experiments on small matrices G. In particular, the following optimization problem was solved numerically:

$$F = \frac{1}{2} \max_{U=U^T \in W_i} \lambda_{\min}\left(UG - \alpha U + (UG - \alpha U)^T\right)$$

where $\lambda_{\min}(A)$ is the smallest eigenvalue of A and α is the least nonzero eigenvalue of G.

By Theorem 3.3, $F \leq 0$. If $F = 0$ for the set of matrices G under consideration, then Conjecture 3.1 is true. For numerical stability in performing this optimization, the set W_i is replaced by the subset $W_{\geq 1} \subset W_i$ which consists of matrices in W_i whose nonzero eigenvalues are larger than or equal to 1. It's clear that if $F = 0$, then using $W_{\geq 1}$ instead of W_i gives the same value for F. Numerical simulations have shown that $F = 0$ for small matrices G which satisfy the properties in Conjecture 3.1. When G does not have a real nonzero eigenvalue whose real part is the smallest among all nonzero eigenvalues, then the value of F returned by the optimization program is strictly less than 0. It remains to be seen what the value of $\mu(G)$ is in this case.

It is clear that a symmetric matrix in W_i is uniquely determined by its superdiagonal elements, i.e., the matrix elements above the main diagonal. Therefore symmetric matrices in W_i can be considered a subset of $R^{\frac{n(n-1)}{2}}$. By the Courant-Fischer theorem F can be written as

$$\begin{aligned}
-F &= \tfrac{1}{2}\min_{U=U^T \in W_i} \lambda_{\max}\left((\alpha U - UG) + (\alpha U - UG)^T\right) \\
&= \min_{U=U^T \in W_i} \max_{\|x\|=1} x^T(\alpha U - UG)x \qquad (3.13)\\
&= \min_{U=U^T \in W_i} h(U)
\end{aligned}$$

where $h(U) = \max_{\|x\|=1} x^T(\alpha U - UG)x$. Next we show that Eq. (3.13) is a *convex programming* problem, i.e. h is a convex function of U and $\mathcal{C} = \{U : U = U^T, U \in W_i\}$ is a convex set. Local minima in convex programming problems are necessarily global minima [68] and efficient algorithms exist for certain classes of convex programming problems [69]. It is easy to see that h is a convex function of U. Suppose $U_1, U_2 \in \mathcal{C}$. Then $U_1, U_2 \in W_i$ and by the Courant-Fischer theorem, $\min_{\|x\|=1, x^T e=0} x^T U_i x > 0$ where $e = (1, \ldots, 1)^T$, $i \in \{1,2\}$. Let $U_c = \eta U_1 + (1-\eta) U_2$ for $0 \leq \eta \leq 1$. Then

$$\min_{\|x\|=1, x^T e=0} x^T U_c x = \min_{\|x\|=1, x^T e=0} x^T \left(\eta U_1 + (1-\eta) U_2\right) x$$
$$\geq \eta \min_{\|x\|=1, x^T e=0} x^T U_1 x + (1-\eta) \min_{\|x\|=1, x^T e=0} x^T U_2 x > 0$$

which implies that $U_c \in W_i$ and since U_c is symmetric, $U_c \in \mathcal{C}$. Therefore $\mathcal{C}$ is a convex set[†].

Example 3.1 Consider Eq. (3.8) where G is the connectivity matrix corresponding to the directed path graph DP_n (Fig. 5.3 in Section 5.1). By Theorem 3.4 $\mu(G) = 1$ and the array synchronizes if there exists a symmetric $V > 0$ such that $f(x,t) + Dx$ is V-uniformly decreasing and VD is symmetric negative semidefinite. In fact, if G is triangular, the requirement that VD is symmetric negative semidefinite is not necessary.

Theorem 3.6 *Suppose G is upper triangular as in Eq. (3.12). If $f(x,t) + a_{i,i} Dx$ is V-uniformly decreasing for each $1 \leq i \leq n-1$ and some symmetric positive definite V, then system (3.8) is synchronizing.*

Proof: We show that $x_i \to x_{i+1}$ by induction. Since $\dot{x}_n = f(x_n, t)$ and $\dot{x}_{n-1} = f(x_{n-1}, t) + a_{n-1,n-1} D x_{n-1} - a_{n-1,n-1} D x_n$, it follows from the discussion in Section 2.2 that $x_{n-1} \to x_n$ as $f(x,t) + a_{n-1,n-1} Dx$ is V-uniformly decreasing. Suppose that $x_j \to x_{j+1}$ for all $j \geq i$. Let $B_1 = \sum_{i \leq j \leq n} a_{i-1,j} D x_j$ and $B_2 = -a_{i-1,i-1} D x_i + \sum_{i \leq j \leq n} a_{i,j} D x_j$. Then $\dot{x}_{i-1} = f(x_{i-1}, t) + a_{i-1,i-1} D x_{i-1} + B_1$ and $\dot{x}_i = f(x_i, t) + a_{i-1,i-1} D x_i + B_2$. Since $\sum_{i \leq j \leq n} a_{i-1,j} = -a_{i-1,i-1} = -a_{i-1,i-1} + \sum_{i \leq j \leq n} a_{i,j}$, it follows that $B_1 \to B_2$. Since $f(x,t) + a_{i-1,i-1} Dx$ is u-asymptotically stable by Corollary C.1, it follows that $x_{i-1} \to x_i$. □

The reader is referred to [13] for more general reducible matrices G which are decomposed into irreducible components.

3.2 Uniform nonlinear static coupling

The coupling in the array described by Eq. (3.8) can be made nonlinear by replacing the operator $G \otimes D$ by a nonlinear operator N. In this case Theorem 3.2 is still valid by a corresponding change to Eq. (3.9): the

---

[†]It can be shown that $\{U : U = U^T, U \in W_{\geq 1}\}$ is also convex.

array synchronizes if the nonlinear operator $(U \otimes V)(N - I \otimes T)$ is negative semidefinite[‡].

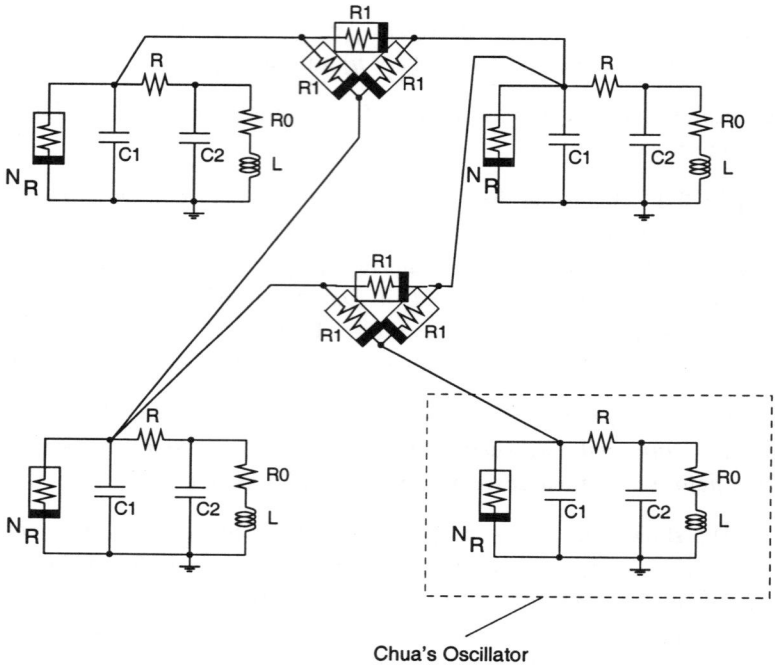

Fig. 3.5 Array of Chua's oscillators coupled with nonlinear resistors R_1.

For example, consider the array of Chua's oscillator coupled by nonlinear resistors shown in Fig. 3.5. The state equations can be written in the form

$$\dot{x} = \begin{pmatrix} f(x_1, t) \\ \vdots \\ f(x_m, t) \end{pmatrix} + (I \otimes D)(M \otimes I)(I \otimes g)(M^T \otimes I)x \qquad (3.14)$$

[‡]A nonlinear operator $L : R^n \to R^n$ is negative semidefinite if $x^T L(x) \leq 0$ for all x.

where f is given by $\hat{f}$ in Eq. (3.7),

$$M = \begin{pmatrix} 1 & 1 & 0 & 0 & 0 & 0 \\ -1 & 0 & 1 & -1 & 0 & 1 \\ 0 & -1 & -1 & 1 & 1 & 0 \\ 0 & 0 & 0 & 0 & -1 & -1 \end{pmatrix}$$

$g(v) = (-g_v(v_1), 0, 0))^T$ and g_v is the v-i characteristic of the nonlinear resistor R_1.

For circuit parameters where all the linear components are passive and the nonlinear resistors N_R have global Lipschitz characteristics, if R_1 is sufficiently passive in the sense that $g_v(v)v \geq Gv^2$ for a large enough G, then the array synchronizes.

More general types of coupling and extensions to hypergraphs can be found in [13] and [70].

3.3 From stability results to synchronization criteria in coupled arrays

We have seen in Chapter 2 that well known stability criteria can be used to derive synchronization criteria in two coupled systems. We will now show that we can also derive synchronization criteria for an array of coupled systems [71, 72]. In this section, we will assume that G is a symmetric matrix in W_i. This implies that all eigenvalues of G are nonnegative.

Consider the coupled array of identical systems with uniform linear coupling (3.8). Application of Corollary 3.4 includes solving the following problem: does there exist a symmetric positive definite P such that $f(x,t) + Dx$ is P-uniformly decreasing and PD is negative semidefinite. There are several stability results in the literature that give conditions under which such a P exists, among them Theorems 2.4, 2.5, 2.7. By applying them to Corollary 3.4, we obtain the following synchronization criteria for the array of coupled systems (3.8).

Theorem 3.7 *Let $f(x,t)$ be of the form $\mu D - B\phi(Cx,t)$ where ϕ is an increasing function. Let D be Hurwitz and let $H(s)$ be defined as $C(sI - D)^{-1}B$. Suppose that (D,B,C) form a minimal realization of $H(s)$. If $H(s)$ is strictly positive real and all nonzero eigenvalues of G are larger than $-\mu$, then the coupled system (3.8) is synchronizing.*

Proof: Let $\alpha > 0$ be the smallest nonzero eigenvalue of G. Note that $G \otimes D = \frac{1}{\delta} G \otimes \delta D$ for a nonzero scalar δ. This trivial observation will be used throughout this section. It's clear that if $C(sI - D)^{-1}B$ is strictly positive real, then $C(sI - (\delta + \mu)D)^{-1}B$ is strictly positive real provided $\delta + \mu > 0$. By setting $\delta = \alpha$, the nonzero eigenvalues of $\frac{1}{\delta}G$ are all larger than or equal to 1 and $\delta + \mu > 0$. By Theorem 2.4 there exists a matrix $P = P^T > 0$ such that $f(x,t) + \delta Dx = (\delta + \mu)x - B\phi(Cx, t)$ is P- uniformly decreasing and $(\mu + \alpha)(PD + D^T P) < 0$ which implies $\alpha(PD + D^T P) < 0$ and Corollary 3.4 can be applied. □

Corollary 3.7 *Let $f(x,t)$ be of the form $-B\phi(Cx,t)$ where ϕ is an increasing function. Let D be Hurwitz and let $H(s)$ be defined as $C(sI - D)^{-1}B$. Suppose that (D, B, C) form a minimal realization of $H(s)$. If $H(s)$ is strictly positive real, then the coupled system (3.8) is synchronizing.*

Theorem 3.8 *Let $f(x,t)$ be of the form $\mu D - b\phi(c^T x, t)$ where ϕ is a function satisfying Eq. (2.11) for some $k > 0$. Let $h(s)$ be the scalar transfer function defined as $c^T(sI - D)^{-1}b$ and suppose that (D, b, c^T) form a minimal realization of $h(s)$. If all nonzero eigenvalues of G are larger than $\max(-\mu, -\mu - k \inf_{\omega \in R} \Re(h(j\omega)))$ and D is Hurwitz, then the coupled system (3.8) synchronizes.*

Proof: Again note that $G \otimes D = \frac{1}{\delta} G \otimes \delta D$. Let $\beta = \inf_{\omega \in R} \Re(h(j\omega))$ and $\alpha > 0$ be the smallest nonzero eigenvalue of G. Choose $\delta = \alpha$. Then $\delta + \mu > 0$, $\delta + \mu > -k\beta$ which implies that $\frac{k\beta}{\delta + \mu} > -1$. All the nonzero eigenvalues of $\frac{1}{\delta}G$ are larger than or equal to 1 and $(\delta + \mu)D$ is Hurwitz. Furthermore, $\Re(1 + kc^T(jwI - (\delta + \mu)D)^{-1}b) \geq 1 + \frac{k\beta}{\delta + \mu} > 0$ for all $w \in R$. By Theorem 2.5 there exists $P = P^T > 0$ such that $f(x,t) + \delta Dx = (\delta + \mu)x - b\phi(c^T x, t)$ is P-uniformly decreasing and $(\alpha + \mu)(PD + D^T P) < 0$ which implies that $\alpha(PD + D^T P) < 0$ and we can apply Corollary 3.4. □

Using Theorem 2.7 and Corollary 3.4, we have the following theorem which describes a synchronization criteria based solely on properties of G, D and the Lipschitz constant γ. Recall that $\sigma_{\max}(A)$ and $\sigma_{\min}(A)$ denote the largest and smallest singular value of A respectively.

Theorem 3.9 *System (3.8) synchronizes if*

- *$f(x,t)$ is Lipschitz constinuous in x for the $\|\cdot\|_2$ norm with Lipschitz constant γ,*

- D is Hurwitz and diagonalizable,
- all nonzero eigenvalues of G are larger than $\frac{\gamma}{\sigma_{\min}(D-j\omega I)}$ for all $\omega \geq 0$.

Proof: Define $\beta = \min_{\omega \geq 0} \sigma_{\min}(D - j\omega I)$. Since D has all its eigenvalues in the open left half plane and diagonalizable, $\beta > 0$ (see Theorem 5 in [22]). Let α be the smallest nonzero eigenvalue of G. Choose $\delta = \alpha > 0$. All singular values of $\delta D - j\omega I$ are greater than $\delta\beta > \gamma$ while the nonzero eigenvalues of $\frac{1}{\delta}G$ are larger than or equal to 1. By substituting $C = -\delta D$, $A = 0$, $L = I$ in Theorem 2.7 we get $P = P^T > 0$ such that $f(x,t) + \delta Dx$ is P-uniformly decreasing and $\alpha(PD + D^T P) = -\gamma^2 PP - \nu I < 0$. □

In [13, 73], it was shown that if D is a diagonal matrix with negative diagonal entries, and f has a bounded Jacobian, then the array (3.8) synchronizes if the nonzero eigenvalues of G are large enough. Theorem 3.9 generalizes this result to all diagonalizable Hurwitz matrices D. Furthermore, Theorem 3.9 gives an explicit bound for the nonzero eigenvalues of G where synchronization occurs.

The following result can also be derived from [22]:

Theorem 3.10 *Let $g(x,t)$ be Lipschitz continuous in x for the $\|\cdot\|_2$ norm with Lipschitz constant γ. If the following conditions are satisfied:*

- *(A, D) is observable,*
- *$A + D$ is Hurwitz,*
- *all singular values of $A + D - j\omega I$ are larger than $\max(\gamma, \sigma_{\max}(A))$ for all $\omega \geq 0$,*

then there exists a symmetric positive definite matrix P such that $Ax + g(x,t) + Dx$ is P-uniformly decreasing and $PD + D^T P$ is negative definite.

Proof: Let $\eta = \max(\gamma, \sigma_{\max}(A))$. In [22] a symmetric positive definite matrix P is constructed such that $(A+D)^T P + P(A+D) + \eta^2 PP + \nu I = 0$, where $\nu > 1$ and $Ax + g(x,t) + Dx$ is P-uniformly decreasing. It then follows that $-(D^T P + PD) = A^T P + PA + \eta^2 PP + \nu I = (\frac{1}{\eta}A + \eta P)^T(\frac{1}{\eta}A + \eta P) + (\nu I - \frac{1}{\eta^2}A^T A)$ which is positive definite if $\sigma_{\max}(A) \leq \eta$. □

This translates into the following synchronization criteria:

Theorem 3.11 *System (3.8) synchronizes if there exists $\mu > 0$ such that*

- *$f(x,t)$ is of the form $Ax + g(x,t)$ where g is Lipschitz continuous in x for the $\|\cdot\|_2$ norm with Lipschitz constant γ,*

- (A, D) is observable,
- $A + \mu D$ is Hurwitz,
- all singular values of $A + \mu D - j\omega I$ are larger than $\max(\gamma, \sigma_{\max}(A))$ for all $\omega \geq 0$,
- all nonzero eigenvalues of G are larger than or equal to μ.

In [22], the test on the singular values of $A + \mu D - j\omega I$ can be replaced by a test on the eigenvalues of $A + \mu D$ and the conditioning of its eigenvectors:

Theorem 3.12 *If all the eigenvalues λ of $A + \mu D$ satisfy*

$$Re(-\lambda) > cond(T)\gamma$$

where T diagonalizes $A + \mu D$ and cond(T) is the condition number of T, then

$$\min_{\omega \geq 0} \sigma_{\min}(A + \mu D - j\omega I) > \gamma$$

3.4 Discrete-time systems

Consider the coupled array of discrete-time systems:

$$\begin{aligned} x_1(k+1) &= f_1(x_1(k), x_1(k), x_2(k), \ldots, x_n(k), k) \\ x_2(k+1) &= f_2(x_2(k), x_1(k), x_2(k), \ldots, x_n(k), k) \\ &\vdots \\ x_n(k+1) &= f_n(x_n(k), x_1(k), x_2(k), \ldots, x_n(k), k) \end{aligned} \quad (3.15)$$

When the individual systems are autonomous (i.e. f does not depend on k) and can be represented as maps they are generally known as coupled map lattices [74]. The corresponding theorem to Theorem 3.1 is:

Theorem 3.13 *The coupled system (3.15) synchronizes if $f_1 = f_2 = \ldots = f_n$ and $x(k+1) = f_1(x(k), \sigma_1(k), \ldots, \sigma_n(k), k)$ is asymptotically stable for all σ_i's.*

Similar to Theorem 3.2, synchronization in an array of coupled discrete-time systems can be deduced from the eigenvalues of the coupling matrix:

Theorem 3.14 *Consider the following array of coupled discrete-time systems:*

$$x(k+1) = \begin{pmatrix} x_1(k+1) \\ \vdots \\ x_n(k+1) \end{pmatrix} = (I - G \otimes D) \begin{pmatrix} f(x_1(k), k) \\ \vdots \\ f(x_n(k), k) \end{pmatrix}$$
$$= (I - G \otimes D) F(x(k), k) \qquad (3.16)$$

where G is a normal $n \times n$ matrix with zero row sums. Let V be a symmetric positive definite matrix such that

$$(f(z,k) - f(\tilde{z},k))^T V (f(z,k) - f(\tilde{z},k)) \leq c^2 (z - \tilde{z})^T V (z - \tilde{z}) \qquad (3.17)$$

for $c > 0$ and all $k, z, \tilde{z}$. If $\|I - \lambda CDC^{-1}\|_2 < \frac{1}{c}$ for every eigenvalue λ of G in $L(G)$, where $V = C^T C$ is the Cholesky decomposition of V and $L(G)$ is defined in Definition 3.8, then the coupled system (3.16) synchronizes, i.e. $x_i(k) \to x_j(k)$ for all i, j as $k \to \infty$.

Proof: First consider the case $V = I$, i.e., $f(x, k)$ is Lipschitz continuous in x with Lipschitz constant c. Since G is normal, G has an orthonormal set of eigenvectors. Denote A as the subspace of vectors of the form $(1, \ldots, 1)^T \otimes v$. Since $G(1, \ldots, 1)^T = 0$ it follows that A is in the kernel of $G \otimes D$. Let b_i be the eigenvectors of G of unit norm corresponding to the eigenvalues λ_i in $L(G)$. Let us denote the subspace orthogonal to A by B. Decompose $x(k)$ as $x(k) = y(k) + z(k)$ where $y(k) \in A$ and $z(k) \in B$. By hypothesis $\|F(x(k), k) - F(y(k), k)\| \leq c\|z(k)\|$. We can decompose $F(x(k), k) - F(y(k), k)$ as $F(x(k), k) - F(y(k), k) = a(k) + b(k)$ where $a(k) \in A$ and $b(k) \in B$. Note that $F(y(k), k)$ and $a(k)$ are in A and therefore $(G \otimes D) F(y(k), k) = (G \otimes D) a(k) = 0$.

$$\begin{aligned} x(k+1) &= (I - G \otimes D) F(x(k), k) \\ &= (I - G \otimes D)(a(k) + b(k) + F(y(k), k)) \\ &= a(k) + F(y(k), k) + (I - G \otimes D) b(k) \end{aligned}$$

Because of the orthogonality of $a(k)$ and $b(k)$, $\|b(k)\| \leq \|F(x(k), k) - F(y(k), k)\| \leq c\|z(k)\|$. Since $b(k) \in B$, it is of the form $b(k) = \sum_i \alpha_i b_i \otimes b_d(k)$.

$$(I - G \otimes D) b(k) = \sum_i \alpha_i b_i \otimes b_d(k) - \sum_i \alpha_i \lambda_i b_i \otimes D b_d(k)$$

$$= \sum_i \alpha_i b_i \otimes (I - \lambda_i D) b_d(k)$$

Therefore

$$\begin{aligned}\|(I - G \otimes D)b(k)\|^2 &= \sum_i |\alpha_i|^2 \|(I - \lambda_i D)b_d(k)\|^2 \\ &< \sum_i |\alpha_i|^2 \frac{\|b_d(k)\|^2}{c^2} = \frac{\|b(k)\|^2}{c^2} \leq \|z(k)\|^2\end{aligned}$$

by hypothesis. Since $a(k) + F(y(k), k)$ is in A, $z(k+1)$ is the orthogonal projection of $(I - G \otimes D)b(k)$ onto B and thus $\|z(k+1)\| \leq \|(I - G \otimes D)b(k)\| < \|z(k)\|$. In fact, $\|z(k+1)\|/\|z(k)\|$ is bounded away from 1 and thus $z(k) \to 0$ as $k \to \infty$. This means that $x(k)$ approaches A, the synchronization manifold.

For general V, let $C^T C$ be the Cholesky decomposition of V. Applying the state transformation $y = (I \otimes C)x$, we get

$$\begin{aligned} y(k+1) &= (I \otimes C)(I - G \otimes D)F((I \otimes C^{-1})y(k), k) \\ &= (I - G \otimes CDC^{-1})(I \otimes C)F((I \otimes C^{-1})y(k), k) \\ &= (I - G \otimes CDC^{-1}) \begin{pmatrix} \tilde{f}(y_1(k), k) \\ \vdots \\ \tilde{f}(y_n(k), k) \end{pmatrix} \end{aligned}$$

where $\tilde{f}(y_i, k) = Cf(C^{-1}y_i, k)$. Then $\tilde{f}$ being Lipschitz continuous with Lipschitz constant c corresponds to Eq. (3.17) and the conclusion follows. □

When D is a normal matrix, Theorem 3.14 can be further simplified.

Theorem 3.15 *Consider the coupled array in Eq. (3.16) where G and D are normal matrices and G has zero row sums. Suppose f is Lipschitz continuous in x with Lipschitz constant c. If $|1 - \lambda\mu| < \frac{1}{c}$ for every eigenvalue λ of G in $L(G)$ and every eigenvalue μ of D then the coupled system (3.16) synchronizes.*

Proof: Since G and D are normal, so is $G \otimes D$. If $b(k) \in B$, by hypothesis $\|(I - G \otimes D)b(k)\| \leq \frac{\|b(k)\|}{c}$. The rest of the proof is similar to Theorem 3.14. □

A graphical interpretation is that the coupled system synchronizes if $L(G)$ lies in the interior of the intersection of circles of radii $\frac{1}{c|\mu_i|}$ centered at $\frac{1}{\mu_i}$ in the complex plane (Fig. 3.6) where μ_i are the eigenvalues of D. A dual interpretation can be obtained by interchanging the roles of $L(G)$ and the eigenvalues of D.

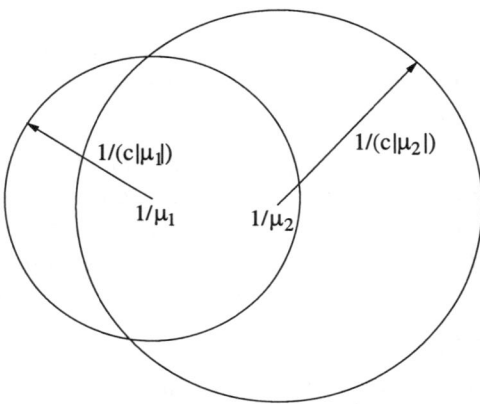

Fig. 3.6 Graphical interpretation of synchronization criterion for arrays of coupled discrete time systems. Coupled array (3.16) synchronizes if all nonzero eigenvalues of G lie inside the intersection of the circles, where μ_i are the eigenvalues of D.

When G and D are symmetric, their eigenvalues are real, and we have the following corollary:

Corollary 3.8 *Let c be the Lipschitz constant of f. If D is symmetric and has only positive eigenvalues between μ_1 and μ_2 (with $0 < \mu_1 \leq \mu_2$) and G is a symmetric matrix with zero row sums and a zero eigenvalue of multiplicity 1 and the nonzero eigenvalues of G are in the interval $\left(\frac{1-\frac{1}{c}}{\mu_1}, \frac{1+\frac{1}{c}}{\mu_2}\right)$ then the coupled system (3.16) synchronizes.*

Note the difference between Corollary 3.4 and Corollary 3.8. In the continuous-time case, the synchronization condition is a condition on the smallest (in magnitude) nonzero eigenvalue of G, while in the discrete-time case, the synchronization condition is a condition on both the smallest and the largest nonzero eigenvalues of G. This is similar to the fact that the stability condition in the continuous time case (eigenvalues in the open left half plane) is mapped to the interior of the unit circle in the discrete time case via the mapping $z \to e^z$.

Consider the special case where the individual systems lie on the nodes of a graph $\mathcal{G}$ and a uniform coupling exists between two systems if and only if there is an edge between the two corresponding nodes. The equation for the i-th system is given by:

$$x_i(k+1) = f(x_i(k),k) - \epsilon \sum_j a_{ij} Df(x_i(k),k) + \epsilon \sum_j a_{ij} Df(x_j(k),k) \quad (3.18)$$

The adjacency matrix and the Laplacian matrix of the graph $\mathcal{G}$ are given by $A = \{a_{ij}\}$ and $V - A$ respectively where V is the diagonal matrix with the vertex degrees on the diagonal.

Corollary 3.9 *Let c be the Lipschitz constant of f. Let the graph $\mathcal{G}$ be connected and let α and β be the largest and the smallest nonzero eigenvalues of the Laplacian matrix $V - A$ of the graph respectively. If D is symmetric and has only positive eigenvalues between μ_1 and μ_2 (with $0 < \mu_1 \leq \mu_2$) and $\frac{\alpha}{\beta} < \frac{(c+1)\mu_1}{(c-1)\mu_2}$ then the system in Eq. (3.18) synchronizes with $\epsilon \in \left(\frac{1-\frac{1}{c}}{\mu_1 \beta}, \frac{1+\frac{1}{c}}{\mu_2 \alpha}\right)$.*

Proof: In this case G is $\epsilon(V - A)$. The largest and smallest nonzero eigenvalues of G is $\epsilon\alpha$ and $\epsilon\beta$ respectively. Furthermore, since $\mathcal{G}$ is connected, the zero eigenvalue of G has multiplicity 1. The condition $\frac{\alpha}{\beta} < \frac{(c+1)\mu_1}{(c-1)\mu_2}$ implies that $\left(\frac{1-\frac{1}{c}}{\mu_1 \beta}, \frac{1+\frac{1}{c}}{\mu_2 \alpha}\right)$ is not empty. Then $\epsilon\beta > \frac{1-\frac{1}{c}}{\mu_1}$ and $\epsilon\alpha < \frac{1+\frac{1}{c}}{\mu_2}$ as needed to apply Corollary 3.8. □

For the special case where D is a positive multiple of the identity matrix, $\mu_1 = \mu_2$ and Corollary 3.9 says that the array in Eq. (3.18) synchronizes for some ϵ if $\frac{\alpha}{\beta}$ is close enough to 1.

Example 3.2 Consider the case where $f(x) = ax(1-x)$ is the logistic map on the interval $[0,1]$ ($0 < a \leq 4$). For $x \notin [0,1]$, we define $f(x) = f(x \bmod 1)$. Then $c = \sup_x |f'(x)| = |a|$. The case of n globally coupled logistic maps can be described by:

$$x_j(i+1) = (1-\epsilon)f(x_j(i)) + \frac{\epsilon}{n}\sum_{k=1}^{n} f(x_k(i))$$

for $j = 1, \ldots, n$ and $\epsilon > 0$. This system is of the form Eq. (3.16) where

$$G = \frac{\epsilon}{n}\begin{pmatrix} n-1 & -1 & -1 & \cdots & -1 \\ -1 & n-1 & -1 & \cdots & -1 \\ \vdots & & \ddots & & \vdots \\ -1 & -1 & \cdots & n-1 & -1 \\ -1 & -1 & \cdots & -1 & n-1 \end{pmatrix}$$

and $D = 1$. The eigenvalues of G are 0 and ϵ. Applying Theorem 3.15 we find that the system will synchronize if $|1 - \epsilon| < \frac{1}{|a|}$.

This bound is somewhat more conservative than that obtained in [75] via numerical simulations. In Fig. 1 of [75], the curve which separates the synchronized state (the region which is denoted "COHERENT" in the figure) from the other states lies below the curve $|1 - \epsilon| = \frac{1}{|a|}$. However, both curves have the general shape with ϵ increasing as a is increased. In [75] it was shown that the local stability condition of the synchronized attractor is given by $|1 - \epsilon| < \frac{1}{e^\lambda}$ where λ is the Lyapunov exponent of the logistic map (see Section 6.2).

Example 3.3 In [76] a lattice of n logistic maps is studied where each map is coupled to k logistic maps randomly chosen among the ensemble. In this case, it was found numerically that the system synchronize for all large enough k. Assume in addition that the coupling is reciprocal, i.e. if there is a coupling between map i and map j, then there is an identical coupling between map j and map i. The corresponding G is then symmetric. There are exactly k nonzero elements in each row of $(I - G)$, each of them equal to $\frac{1}{k}$. Computer simulations indicate that for large n, G has one zero eigenvalue, and almost always the remaining eigenvalues lies in the range $\left[1 - \frac{2}{\sqrt{k}}, 1 + \frac{2}{\sqrt{k}}\right]$. In fact, a result of J. Friedman [77] shows that for large n and even k this is true. We will discuss this in more detail in Chapter 5.

Chapter 4

Synchronization in Coupled Arrays: Dynamic Coupling

In dynamic coupling, the coupling elements can contain dynamic circuit elements and can have dynamics of their own. In this case, we distinguish two types of systems, the *cell* system and the *coupling* system. The cell systems are the systems to be synchronized and we assume that they are identical. The coupling systems are the systems which connects the cells together and we assume that *the cell systems are not coupled to each other directly and similarly for the coupling systems*. In other words, the coupling systems are used to couple several cell systems (Fig. 4.1).

For example, consider the array of Chua's oscillators coupled via relaxation oscillators shown in Fig. 4.2. Such an array of cell systems coupled via the coupling systems is said to *synchronize* if the states of the *cell* systems converge to each other.

In some cases, Theorem 3.1 can still be applied. For example, consider the array of Chua's oscillators coupled via dynamic coupling elements as shown in Fig. 4.3. By replacing the one-ports indicated by P_1 and P_2 with a corresponding time-varying voltage source and applying the Substitution Theorem [6], the state equations can be written as Eq. (3.5) and for linear passive elements and globally Lipschitz N_R, the array synchronizes when the resistance of R_1 is small enough.

In more general cases, assume that the state equations can be written

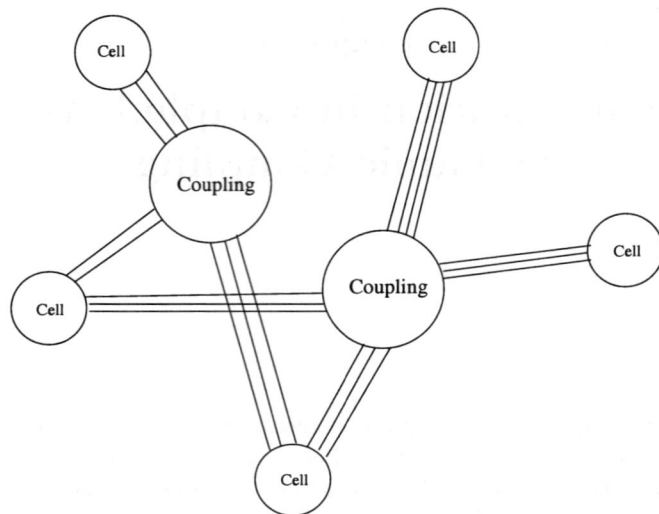

Fig. 4.1 Array of coupled circuits and systems where the cell systems are coupled only via the coupling systems.

in the form:

$$\begin{aligned}
\dot{x}_1 &= f_1(x_1, y_1, y_2, \ldots, y_m, t) \\
&\vdots \\
\dot{x}_n &= f_n(x_n, y_1, y_2, \ldots, y_m, t) \\
\dot{y}_1 &= g_1(y_1, x_1, x_2, \ldots, x_n, t) \\
&\vdots \\
\dot{y}_m &= g_m(y_m, x_1, x_2, \ldots, x_n, t)
\end{aligned} \qquad (4.1)$$

where x_i are the states of the cell systems and y_i are the states of the coupling systems.

Under certain conditions, we can define a connectivity hypergraph for Eq. (4.1).

Definition 4.1 Suppose that in Eq. (4.1), if f_i depends on y_j, then g_j depends on x_i. The *connectivity hypergraph* of coupled array (4.1) is defined as the hypergraph with vertex set $V = \{v_i\}$ and edge set E defined by

$\{v_{i_1}, \ldots v_{i_k}\} \in E \Leftrightarrow$ there exists j such that g_j depends only on $x_{i_1}, \ldots x_{i_k}$

The vertices of the connectivity hypergraph correspond to the cell sys-

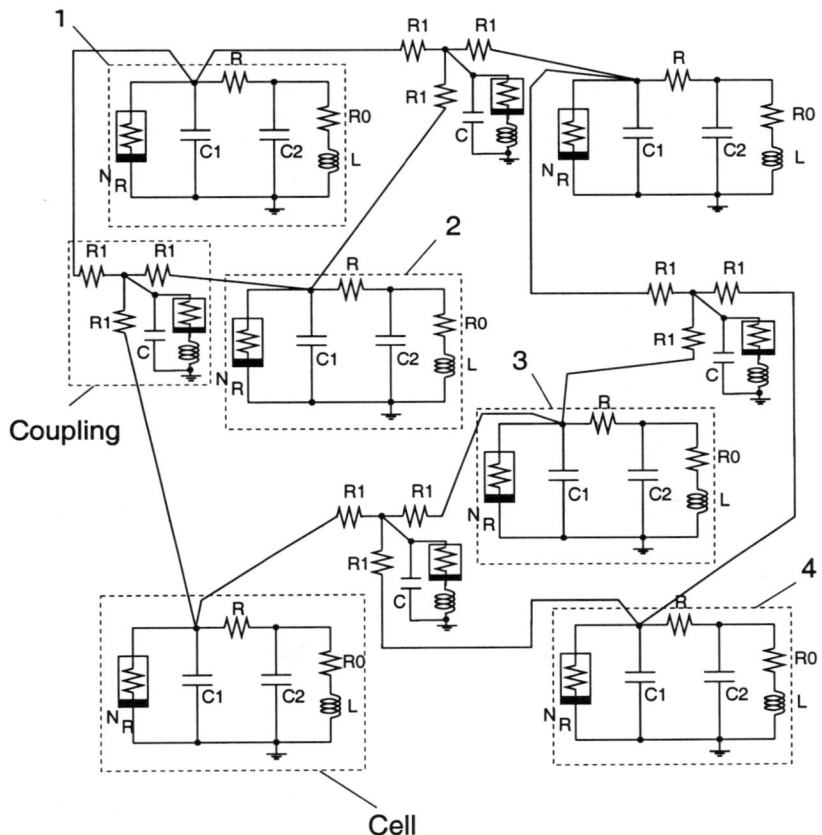

Fig. 4.2 Array of Chua's oscillators (cell) coupled linearly via nonlinear oscillators (coupling).

tems and the edges of the connectivity hypergraph correspond to the coupling systems. For example the hypergraph corresponding to Fig. 4.2 is shown in Fig. 4.4 with incidence matrix

$$\begin{pmatrix} 1 & 1 & 0 & 0 \\ 1 & 0 & 1 & 0 \\ 1 & 1 & 0 & 0 \\ 0 & 0 & 1 & 1 \\ 0 & 1 & 0 & 1 \\ 0 & 0 & 1 & 1 \end{pmatrix} \quad (4.2)$$

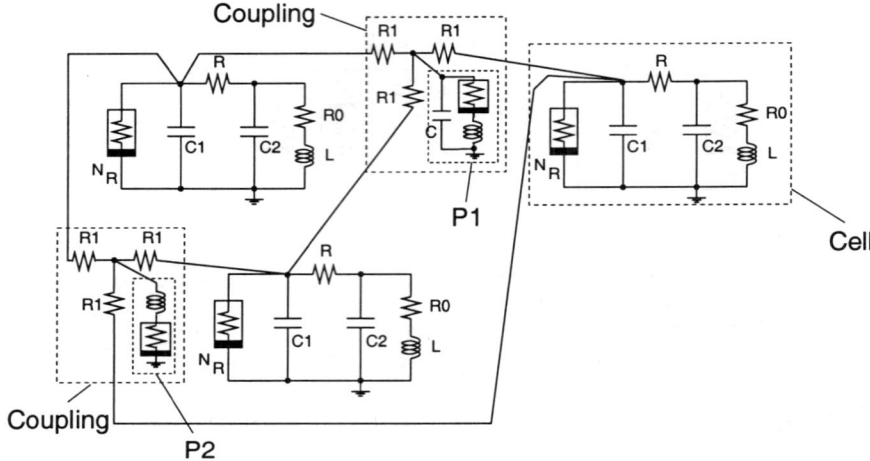

Fig. 4.3 Array of Chua's oscillators coupled linearly via nonlinear oscillators for which Theorem 3.1 can be applied.

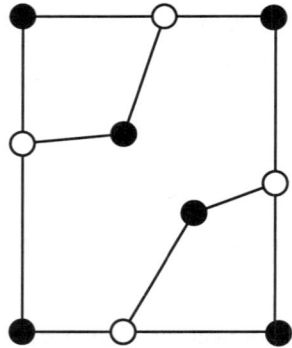

Fig. 4.4 Hypergraph corresponding to the coupled array in Fig. 4.2. The closed circles represent vertices while the open circles represent edges.

The roles of cell systems and coupling systems can be interchanged (i.e. changing x_i into y_i in Eq. (4.1) and vice versa) resulting in a *dual* coupled system. The connectivity hypergraph of the dual coupled system is the dual hypergraph.

An analog of Theorem 3.1 is as follows:

Theorem 4.1 *The coupled system*

$$\begin{aligned}
\dot{x}_1 &= f_1(x_1, y_{s_{11}}, y_{s_{12}}, \ldots, y_{s_{1k}}, t) \\
&\vdots \\
\dot{x}_n &= f_n(x_n, y_{s_{n1}}, y_{s_{n2}}, \ldots, y_{s_{nk}}, t) \\
\dot{y}_1 &= g_1(y_1, x_{t_1}, \ldots, x_{t_l}, t) \\
&\vdots \\
\dot{y}_m &= g_m(y_m, x_{t_1}, \ldots, x_{t_l}, t)
\end{aligned} \quad (4.3)$$

where $s_{ij} \in \{1,\ldots,m\}$ and $t_i \in \{1,\ldots,n\}$ synchronizes in the sense that $|x_i(t) - x_j(t)| \to 0$ as $t \to \infty$ if

(1) $f_1 = f_2 = \ldots = f_n$,
(2) $\dot{x} = f_1(x, \sigma_1(t), \ldots, \sigma_k(t), t)$ *is u-asymptotically stable for all σ_i's,*
(3) $g_1 = g_2 = \ldots = g_m$,
(4) $\dot{y} = g_1(y, \eta_1(t), \ldots, \eta_l(t), t)$ *is asymptotically stable for all η_i's.*

The proof follows directly from the definitions of asymptotical and u-asymptotical stability. If the conditions of Theorem 4.1 are satisfied, not only are the cell systems' trajectories converging to each other, but the coupling systems' trajectories are also converging to each other, i.e., $|y_i(t) - y_j(t)| \to 0$ as $t \to \infty$.

For example, consider the coupled array of Chua's oscillators coupled with relaxation oscillators in Fig. 4.5. The state equations can be written in the form (4.3). For passive linear circuit elements and (active) nonlinear resistors which are globally Lipschitz, it can be shown that for a small enough R_1, the conditions of Theorem 4.1 are satisfied and thus the array synchronizes.

4.1 Synchronization of clusters

So far we have discussed synchronization where *all* the circuits are identical and they all synchronize to each other. In some numerical studies of coupled systems, the object of interest is when they are not all synchronized, but the circuits form groups where circuits within a group synchronize to each other, and synchronization does not necessary occur between circuits from different groups. This partitioning into groups can change with time and the

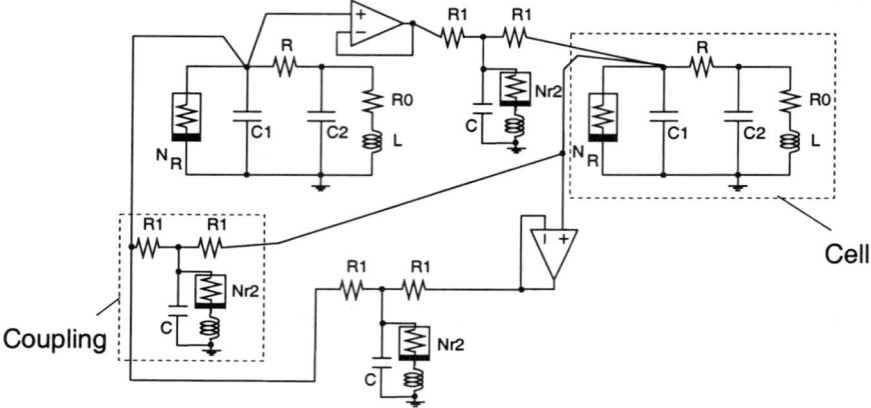

Fig. 4.5 An array of Chua's oscillators coupled via relaxation oscillators to illustrate the application of Theorem 4.1.

coupled systems can form spatio-temporal patterns and wave phenomena [78, 79].

We call synchronization where the circuits are partitioned into groups and circuits within groups synchronize *clustered synchronization*. The synchronization of arrays with dynamic coupling in Theorem 4.1 is a special case of clustered synchronization with 2 clusters: a cluster consisting of circuits and a cluster consisting of the coupling elements.

Definition 4.2 An $n \times m$ matrix T is in class $\mathcal{T}$ if it contains only nonnegative entries and all its nonzero entries lie on the main diagonal.

Definition 4.3 For an $n \times m$ matrix A, $\phi(A)$ is a diagonal $n \times n$ matrix, where the diagonal entries are the row sums of A.

Lemma 4.1 *If T is in class $\mathcal{T}$, then $\phi(T)\phi^\dagger(T)T = T$ and $T\phi^\dagger(T^T)T^T = \phi(T)$, where $\phi^\dagger(T)$ is the pseudo-inverse (Moore-Penrose inverse) of $\phi(T)$.*

The proof of this Lemma is trivial.

Consider the state equations of the coupled array written in the following form:

$$\begin{aligned}\dot{x} &= I_n \otimes g(x_i) + (E \otimes T_1(y_i)) - (\phi(E) \otimes T_2(x_i)) \\ \dot{y} &= \hat{h}(x,y)\end{aligned} \quad (4.4)$$

where the x_i's are the state vectors of the cells and y_i's are the state vectors of the coupling elements and

$$x = \begin{pmatrix} x_1 \\ \vdots \\ x_n \end{pmatrix}, \quad y = \begin{pmatrix} y_1 \\ \vdots \\ y_m \end{pmatrix}.$$

The vector field g determines the dynamics of the uncoupled cells and $\hat{h}$ is the vector field governing the coupling elements.

The following result gives conditions under which some of the oscillators are synchronized to each other.

Theorem 4.2 *If the j-th row of E is equal to the k-th row of E and $g - \phi(E)_{j,j}T_2$ is uniformly decreasing, then $x_j \to x_k$ as $t \to \infty$, i.e. oscillator j is synchronized with oscillator k.*

Proof: It is clear that $\phi(E)_{j,j} = \phi(E)_{k,k}$. Let E_i denote the i-th row of E. Since $\dot{x}_j = g(x_j) + E_j \otimes T_1(y_i) - \phi(E)_{j,j}T_2(x_j)$ and $\dot{x}_k = g(x_k) + E_k \otimes T_1(y_i) - \phi(E)_{k,k}T_2(x_k)$, the result follows from Theorem 2.1. □

Example 4.1 Consider Chua's oscillators coupled via coupling elements consisting of nonlinear oscillators as shown in Fig. 4.6.

The state equations can be written in the form of Eq. (4.4) if we choose $x_i = (V_{1i}, V_{2i}, I_{3i})^T$ as the state of the Chua's oscillator and $y_i = (V_{Ci}, I_{Li})^T$ as the state of the coupling and $g(x_i) = \begin{pmatrix} \frac{1}{C_1}\left(\frac{V_{2i}-V_{1i}}{R_1} - f(V_{1i})\right) \\ \frac{1}{C_2}\left(\frac{V_{1i}-V_{2i}}{R_1} + I_{3i}\right) \\ -\frac{1}{L}(V_{2i} + R_0 I_{3i}) \end{pmatrix}$ where $f(V)$ is the v-i characteristic of the nonlinear resistor N_R. Furthermore, $T_2(x_i) = (\frac{V_{1i}}{R_1}, 0, 0)^T$ and $\hat{h}(x,y) = I \otimes h(y_i) + E^T \otimes (\frac{V_{1i}}{R_1}, 0)^T + \phi(E^T) \otimes (\frac{V_{Ci}}{R_1}, 0)^T$ where $h(y_i) = \begin{pmatrix} \frac{I_{Li}}{C} \\ -\frac{1}{L_2}(V_{Ci} + N_{R2}(I_{Li})) \end{pmatrix}$. We assume that all linear elements in Chua's oscillators are passive and the nonlinear resistor N_R is globally Lipschitz.

The incidence matrix E of the hypergraph is given by Eq. (4.2). The first and third row of E are equal and the fourth and sixth row of E are equal. By using Theorem 4.2 it can be shown that Chua's oscillator 1 synchronizes to oscillator 2 and Chua's oscillator 2 synchronizes to Chua's oscillator 4 for small enough $R_1 > 0$. Fig. 4.7 shows time waveforms of V_{2i}

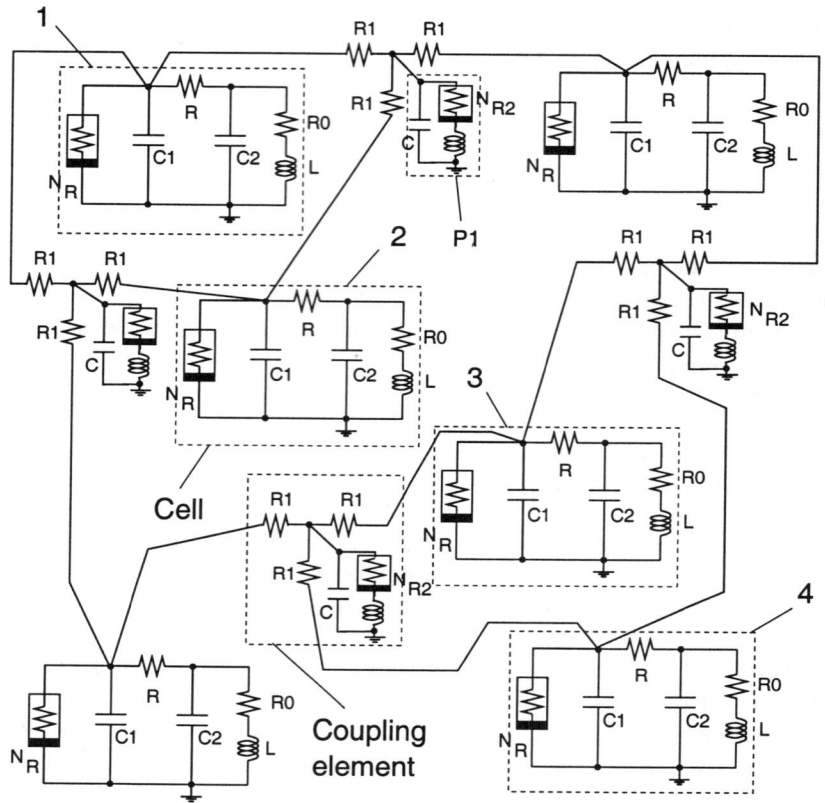

Fig. 4.6 Array of Chua's oscillators coupled via dynamic coupling elements.

of the Chua's oscillators. Note that all the oscillators are almost but not exactly synchronized to each other.

Let us look at the synchronization from a circuit theoretic viewpoint. By the Substitution Theorem [6], we can replace the one port in each coupling element (denoted by p_1) by a voltage source. It is then clear that oscillator 1 and oscillator 2 are driven by the same voltage sources V_1 and V_2 coupled via R_1 as shown in Fig. 4.8. This is equivalent via a Thévenin-Norton transformation to Fig. 4.9 with $R_2 = \frac{1}{2}R_1$. Since a small R_2 in parallel with N_R is passive, this is equivalent to two passive circuits driven by the same current source and thus their states converge towards each other.

Theorem 3.1 can be generalized to include clustered synchronization:

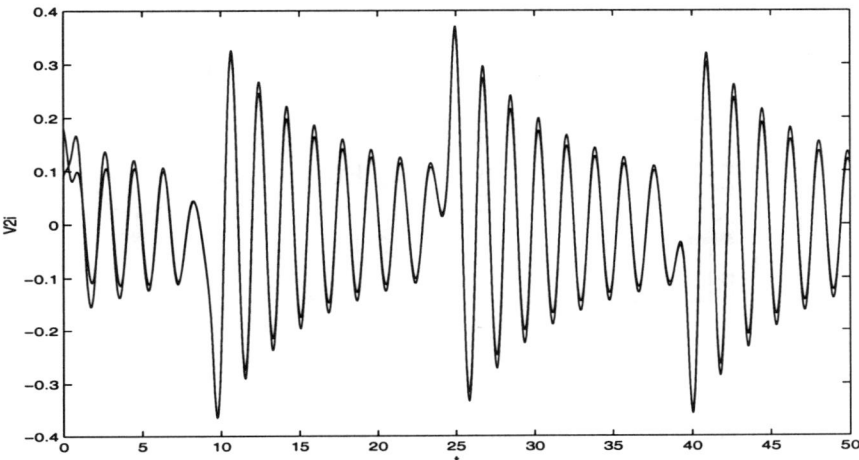

Fig. 4.7 Time waveforms of V_{2i} in an array of Chua's oscillators coupled via dynamic coupling elements. Note that the oscillators are almost synchronized to each other.

Theorem 4.3 *Consider the coupled array of circuits written in the form:*

$$
\begin{aligned}
\dot{x}_{11} &= f_1(x_{11}, x_{11}, \ldots, x_{kl}, \ldots, t) \\
\dot{x}_{12} &= f_1(x_{12}, x_{11}, \ldots, x_{kl}, \ldots, t) \\
&\vdots \\
\dot{x}_{1,n_1} &= f_1(x_{1,n_1}, x_{11}, \ldots, x_{kl}, \ldots, t) \\
\dot{x}_{21} &= f_2(x_{21}, x_{11}, \ldots, x_{kl}, \ldots, t) \\
\dot{x}_{22} &= f_2(x_{22}, x_{11}, \ldots, x_{kl}, \ldots, t) \\
&\vdots \\
\dot{x}_{2,n_2} &= f_2(x_{2,n_2}, x_{11}, \ldots, x_{kl}, \ldots, t) \\
&\vdots \\
\dot{x}_{i1} &= f_i(x_{i1}, x_{11}, \ldots, x_{kl}, \ldots, t) \\
&\vdots \\
\dot{x}_{i,n_i} &= f_i(x_{i,n_i}, x_{11}, \ldots, x_{kl}, \ldots, t) \\
&\vdots
\end{aligned}
\qquad (4.5)
$$

If $\dot{x} = f_i(x, \sigma_1, \sigma_2, \ldots, t)$ is asymptotically stable for all i and σ_j, then $x_{ij} \to x_{ik}$ as $t \to \infty$.

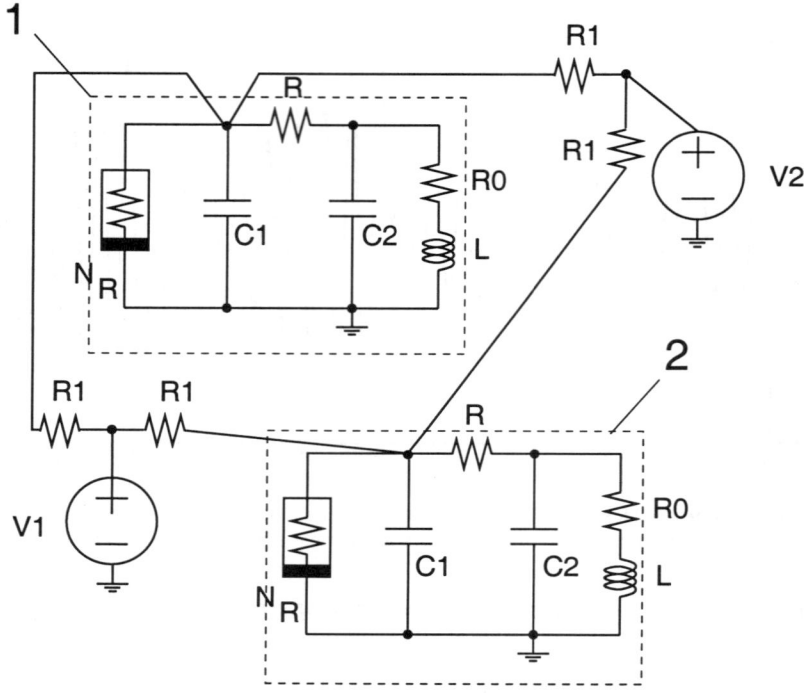

Fig. 4.8 Applying the substitution theorem to replace one-ports in the coupling elements with appropriate voltage sources. Only circuit elements connected to oscillator 1 and oscillator 2 are shown.

What the above theorem says is that the circuits are partitioned into clusters $S_i = \{x_{i1}, \ldots, x_{i,n_i}\}$ and circuits within each cluster synchronize. For instance, Chua's oscillator 1 synchronizes with Chua's oscillator 2 and Chua's oscillator 3 synchronizes with Chua's oscillator 4 in Fig. 4.2 for small enough R_1 when N_R has a globally Lipschitz characteristic and all the linear elements are passive.

A generalization of Theorem 4.1 to this case is also possible:

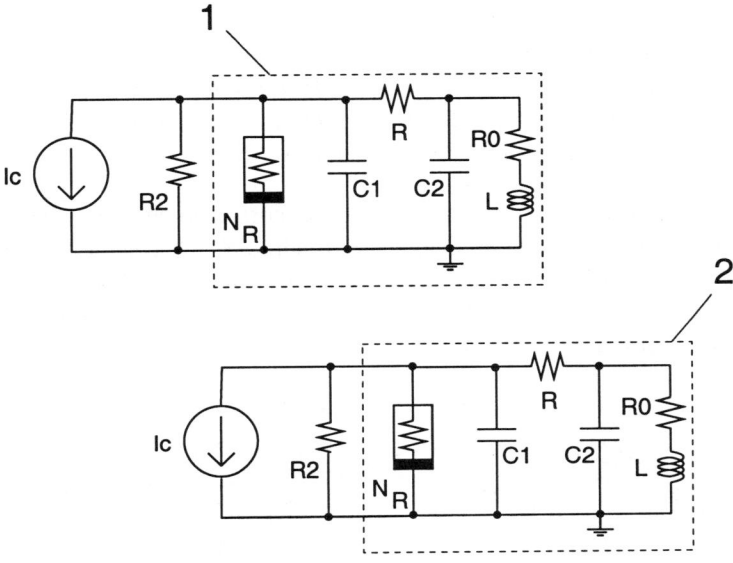

Fig. 4.9 Replacing the voltage sources V_1 and V_2 and resistors in Fig. 4.8 by a Norton equivalent results in two identical passive circuits driven by the same current source I_c.

Theorem 4.4 *Consider the coupled array of circuits written in the form:*

$$
\begin{aligned}
\dot{x}_1 &= f_1(x_1, x_1, \ldots, x_{n_1}, t) \\
\dot{x}_2 &= f_1(x_2, x_1, \ldots, x_{n_1}, t) \\
&\vdots \\
\dot{x}_{n_1} &= f_1(x_{n_1}, x_1, \ldots, x_{n_1}, t) \\
\dot{x}_{n_1+1} &= f_2(x_{n_1+1}, x_{s_1^{n_1+1}}, \ldots, x_{s_{k_2}^{n_1+1}}, t) \\
&\vdots \\
\dot{x}_{n_2} &= f_2(x_{n_2}, x_{s_1^{n_2}}, \ldots, x_{s_{k_2}^{n_2}}, t) \\
&\vdots \\
\dot{x}_{n_{m-1}} &= f_{m-1}(x_{n_{m-1}}, x_{s_1^{n_{m-1}}}, \ldots, x_{s_{k_{m-1}}^{n_{m-1}}}, t) \\
\dot{x}_{n_{m-1}+1} &= f_m(x_{n_{m-1}+1}, x_{s_1^{n_{m-1}+1}}, \ldots, x_{s_{k_m}^{n_{m-1}+1}}, t) \\
&\vdots \\
\dot{x}_{n_m} &= f_m(x_{n_m}, x_{s_1^{n_m}}, \ldots, x_{s_{k_m}^{n_m}}, t)
\end{aligned}
\qquad (4.6)
$$

where $0 = n_0 < n_1 < n_2 < \ldots < n_m$ and $S_l = \{n_l + 1, \ldots n_{l+1}\}$. The indices s_j^i are chosen such that $s_j^i \in \{1, \ldots, n_l\}$ if $i \in S_l$.

If $\dot{x} = f_1(x, \sigma_1, \sigma_2, \ldots, t)$ is asymptotically stable for all σ_i and $\dot{x} = f_k(x, \eta_1, \eta_2, \ldots, t)$ is u-asymptotically stable for all η_i and all $k > 1$, then $x_a \to x_b$ as $t \to \infty$ for a and b in the same set S_l.

The synchronization in the above theorem occurs in m clusters with indices of the states in each cluster belonging to some S_l. A schematic diagram of how the clusters are connected is shown in Fig. 4.10. There is an edge from S_i to S_j only if $i < j$. This means that oscillators in S_j receive input from oscillators in S_i only if $i < j$.

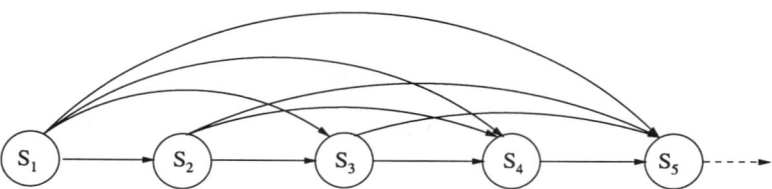

Fig. 4.10 Schematic diagram of the connectivity in Eq. (4.6). There is an edge from S_i to S_j only if $i < j$.

4.2 Regular and uniform hypergraphs in linearly coupled arrays

In this section we consider linearly coupled arrays where the underlying hypergraph is regular and uniform. In this case the Laplacian matrix L can be written as $2(\alpha I - \frac{1}{\beta} E E^T)$ for α and β positive integers representing the vertex and edge degrees respectively. By Lemma A.4 the algebraic connectivity is smaller than or equal to $\frac{2n}{n-1}\left(\alpha - \frac{\alpha}{\beta}\right)$ where n is the number of vertices. Since $\beta \leq n$, this implies that the algebraic connectivity is less than or equal to 2α. Similarly, the algebraic connectivity of the dual hypergraph is less than or equal to 2β. We consider state equations in the general form given by:

$$\dot{x} = I_n \otimes g(x_i) + (E \otimes \gamma_1 T)y - (\alpha I \otimes \gamma_1 \phi(T))x \qquad (4.7)$$
$$\dot{y} = I_m \otimes h(y_i) + (E^T \otimes \gamma_2 T^T)x - (\beta I \otimes \gamma_2 \phi(T^T))y \qquad (4.8)$$

where E is the incidence matrix of the underlying regular and uniform hypergraph and x_i and y_i are the states of the cells and the coupling elements respectively. The scalars $\gamma_1 > 0$, $\gamma_2 > 0$ indicate the coupling strengths and T is a matrix in class $\mathcal{T}$. The following result gives conditions when this coupled array synchronizes [70, 80]:

Theorem 4.5 *The coupled array in Eqs. (4.7-4.8) synchronizes in the sense that $x_i \to x_j$, $y_i \to y_j$, as $t \to \infty$ for all i,j if there exists constants δ_1 and δ_2 such that $g - \delta_1 \phi(T)$ and $h - \delta_2 \phi(T^T)$ are uniformly decreasing and there exists real numbers $a_1 > 0$ and $a_2 > 0$ such that $\sigma_1 \geq 2\left(\frac{4\alpha a_2((a_1-\alpha)\gamma_1 + \delta_1)}{\gamma_2 \beta^2} + \alpha\right)$ and $\sigma_2 \geq 2\left(\frac{4\beta a_1((a_2-\beta)\gamma_2 + \delta_2)}{\gamma_1 \alpha^2} + \beta\right)$ where $\sigma_1 > 0$ and $\sigma_2 > 0$ are the algebraic connectivities of the hypergraph* and its dual respectively.*

Proof: Let L_x and L_y be the Laplacian matrices of the hypergraph and its dual respectively, i.e., $L_x = 2(\alpha I_n - \frac{1}{\beta}EE^T)$ and $L_y = 2(\beta I_m - \frac{1}{\alpha}E^T E)$. By Lemmas B.1 L_x and L_y are matrices in W_i and by Lemma A.12 they can be decomposed as $L_x = M_x^T M_x$ and $L_y = M_y^T M_y$ where M_x and M_y are matrices in M_2. Construct the Lyapunov function:

$$V = \frac{1}{2}\left(x^T(L_x \otimes I)x + y^T(L_y \otimes I)y\right)$$

Its derivative along trajectories is given by:

$$\begin{aligned}\dot{V} &= x^T(L_x \otimes I)\left(I_n \otimes g(x_i) + (E \otimes \gamma_1 T)y - (\alpha I \otimes \gamma_1 \phi(T))x\right) \\ &\quad + y^T(L_y \otimes I)\left(I_m \otimes h(y_i) + (E^T \otimes \gamma_2 T^T)x - (\beta I \otimes \gamma_2 \phi(T^T))y\right) \\ &= x^T(L_x \otimes I)((\delta_1 - \alpha \gamma_1)I \otimes \phi(T))x + x^T(L_x \otimes I)(E \otimes \gamma_1 T)y \\ &\quad + y^T(L_y \otimes I)((\delta_2 - \beta \gamma_2)I \otimes \phi(T^T))y \\ &\quad + y^T(L_y \otimes I)(E^T \otimes \gamma_2 T^T)x + p_1(x) + p_2(y)\end{aligned}$$

where $p_1(x) = x^T(L_x \otimes I)(I_n \otimes g(x_i) - \delta_1(I \otimes \phi(T)))x$ and $p_2(y) = y^T(L_y \otimes I)(I_m \otimes h(y_i) - \delta_2(I \otimes \phi(T^T)))y$ are nonpositive by hypothesis. Furthermore, $p_1(x)$ and $p_2(y)$ are zero if and only if $x_i = x_j$ and $y_i = y_j$ for all i, j respectively. After some further manipulation and application of Lemma

---

*Which is assumed to be connected.

4.1 we get:

$$\dot{V} = p_1(x) + p_2(y)$$
$$-\left\|\sqrt{a_1}(M_x \otimes \sqrt{\phi(\gamma_1 T)})x - \tfrac{1}{2\sqrt{a_1}}(M_x E \otimes \sqrt{\phi(\gamma_1 T)}\phi^\dagger(\gamma_1 T)\gamma_1 T)y\right\|^2$$
$$-\left\|\sqrt{a_2}(M_y \otimes \sqrt{\phi(\gamma_2 T^T)})y - \tfrac{1}{2\sqrt{a_2}}(M_y E^T \otimes \sqrt{\phi(\gamma_2 T^T)}\phi^\dagger(\gamma_2 T^T)\gamma_2 T^T)x\right\|^2$$
$$+b_1 x^T (L_x \otimes \phi(\gamma_1 T))x$$
$$+\tfrac{1}{4a_1} y^T \left(E^T L_x E \otimes \gamma_1 T^T \phi^\dagger(\gamma_1 T)\phi(\gamma_1 T)\phi^\dagger(\gamma_1 T)\gamma_1 T\right) y$$
$$+b_2 y^T (L_y \otimes \phi(\gamma_2 T^T))y$$
$$+\tfrac{1}{4a_2} x^T \left(E L_y E^T \otimes \gamma_2 T \phi^\dagger(\gamma_2 T^T)\phi(\gamma_2 T^T)\phi^\dagger(\gamma_2 T^T)\gamma_2 T^T\right) x$$

where $\sqrt{\phi(\gamma T)}$ is the diagonal matrix whose square is equal to $\phi(\gamma T)$, $b_1 = a_1 + \frac{\delta_1}{\gamma_1} - \alpha$ and $b_2 = a_2 + \frac{\delta_2}{\gamma_2} - \beta$. This simplifies to:

$$\dot{V} = p_1(x) + p_2(y)$$
$$-\left\|\sqrt{a_1}(M_x \otimes \sqrt{\phi(\gamma_1 T)})x - \tfrac{1}{2\sqrt{a_1}}(M_x E \otimes \sqrt{\phi(\gamma_1 T)}\phi^\dagger(\gamma_1 T)\gamma_1 T)y\right\|^2$$
$$-\left\|\sqrt{a_2}(M_y \otimes \sqrt{\phi(\gamma_2 T^T)})y - \tfrac{1}{2\sqrt{a_2}}(M_y E^T \otimes \sqrt{\phi(\gamma_2 T^T)}\phi^\dagger(\gamma_2 T^T)\gamma_2 T^T)x\right\|^2$$
$$+x^T \left((\gamma_1 b_1 L_x + \tfrac{\gamma_2}{4a_2} E L_y E^T) \otimes \phi(T)\right) x$$
$$+y^T \left((\gamma_2 b_2 L_y + \tfrac{\gamma_1}{4a_1} E^T L_x E) \otimes \phi(T^T)\right) y$$
(4.9)

Note that

$$\gamma_1 b_1 L_x + \frac{\gamma_2}{4a_2} E L_y E^T = 2\left(\gamma_1 b_1 \alpha I_n + \left(\frac{-\gamma_1 b_1}{\beta} + \frac{\gamma_2 \beta}{4a_2}\right) E E^T - \frac{\gamma_2}{4a_2 \alpha}(E E^T)^2\right).$$

Since the eigenvalues of EE^T and L_x are related by $\lambda(EE^T) = \beta(\alpha - \frac{\lambda(L_x)}{2})$, the eigenvalues of $\gamma_1 b_1 L_x + \frac{\gamma_2}{4a_2} E L_y E^T$ are given by:

$$2\left(\gamma_1 b_1 \alpha + \beta\left(\frac{-\gamma_1 b_1}{\beta} + \frac{\gamma_2 \beta}{4a_2}\right)\left(\alpha - \frac{\lambda(L_x)}{2}\right) - \frac{\gamma_2 \beta^2}{4a_2 \alpha}\left(\alpha - \frac{\lambda(L_x)}{2}\right)^2\right)$$

This can be simplified to

$$\lambda(L_x)\left(\gamma_1 b_1 + \frac{\gamma_2 \beta^2}{4a_2} - \frac{\gamma_2 \beta^2}{8a_2 \alpha}\lambda(L_x)\right)$$

Since all eigenvalues of L_x are nonnegative, this means that $\gamma_1 b_1 L_x + \frac{\gamma_2}{4a_2} E L_y E^T$ is negative semidefinite if and only if the least positive eigenvalue of L_x is larger than or equal to $2\left(\frac{4\gamma_1 b_1 \alpha a_2}{\gamma_2 \beta^2} + \alpha\right)$, i.e. if and only if

$\sigma_1 \geq 2\left(\frac{4\gamma_1 b_1 \alpha a_2}{\gamma_2 \beta^2} + \alpha\right)$. Similarly, we can show that $\gamma_2 b_2 L_y + \frac{\gamma_1}{4a_1}E^T L_x E$ is negative semidefinite if and only if $\sigma_2 \geq 2\left(\frac{4\gamma_2 b_2 \beta a_1}{\gamma_1 \alpha^2} + \beta\right)$. Since $\phi(T)$ and $\phi(T^T)$ are positive semidefinite, all the terms in Eq. (4.9) are nonpositive and the conclusion follows from Theorem C.4. □

This result suggests that under certain conditions the higher the algebraic connectivities of the hypergraph and its dual, the smaller γ_1 and γ_2 need to be for synchronization, making the array easier to synchronize. This makes sense intuitively as the algebraic connectivity measures how connected the hypergraph is and the more connected the array, the easier it should be to synchronize.

Example 4.2 Consider the array of coupled Chua's oscillators shown in Fig. 4.11. This can be written in the form of Eqs. (4.7-4.8) if we choose g and h as in Example 4.1, $T = \begin{bmatrix} 1 & 0 \\ 0 & 0 \\ 0 & 0 \end{bmatrix}$, $\gamma_1 = \frac{1}{C_1 R_1}$, and $\gamma_2 = \frac{1}{CR_1}$. The underlying hypergraph (Fig. 4.12) is regular and uniform (and is in fact self-dual) and has incidence matrix

$$\begin{bmatrix} 0 & 1 & 1 & 1 \\ 1 & 0 & 1 & 1 \\ 1 & 1 & 0 & 1 \\ 1 & 1 & 1 & 0 \end{bmatrix}$$

The algebraic connectivity of the hypergraph and its dual are both $5\frac{1}{3}$. Again we assume that the linear circuit elements are passive and the nonlinear resistor is globally Lipschitz. Suppose that the nonlinear resistor N_R satisfies $|f(v)| \leq G_N v$ for some $G_N > 0$. This implies that δ_1 can be chosen to be any positive number larger than G_N/C_1. Since N_{R2} is passive, δ_2 can be chosen to be arbitrarily small and positive. It can be shown that for small enough $R_1 > 0$ and $\frac{C}{C_1}$ and for a passive N_{R2}, the conditions of Theorem 4.5 are satisfied and the array synchronizes. For example, by choosing $a_1 = a_2 = 1$, the array synchronizes if $0 < R_1 < \frac{2}{G_N}$ and $0 < C < 4C_1$.

Corollary 4.1 *If the hypergraph is regular and uniform and every edge is the full set of vertices†, then the coupled array in Eqs. (4.7-4.8) synchro-*

---
†i.e., α and β are the number of edges and vertices respectively.

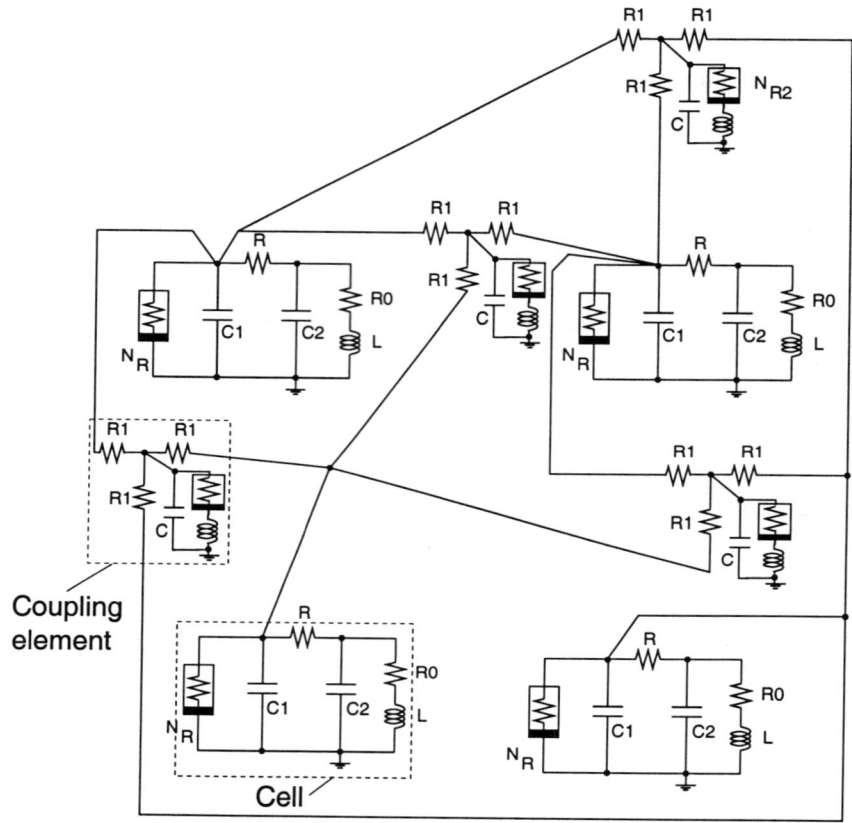

Fig. 4.11 Array of Chua's oscillators coupled via dynamic coupling elements on a regular and uniform hypergraph.

nizes if there exists constants δ_1 and δ_2 such that $g-\delta_1\phi(T)$ and $h-\delta_2\phi(T^T)$ are uniformly decreasing, $\frac{\delta_1}{\gamma_1} < \alpha$, and $\frac{\delta_2}{\gamma_2} < \beta$.

Proof: For this graph $\sigma_1 = 2\alpha$ and $\sigma_2 = 2\beta$ and the inequalities in Theorem 4.5 are satisfied with sufficiently small a_1 and a_2. □

The following theorem give synchronization criteria for the optimal choice of a_1 and a_2.

Theorem 4.6 *Suppose that $\frac{\delta_1}{\gamma_1} < \alpha$, $\frac{\delta_2}{\gamma_2} < \beta$, $\sigma_1 < 2\alpha$, and $\sigma_2 < 2\beta$ where $\sigma_1 > 0$ and $\sigma_2 > 0$ are the algebraic connectivities of the hypergraph and its dual respectively.. The coupled array in Eqs. (4.7-4.8) synchronizes*

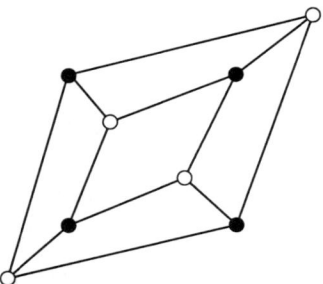

Fig. 4.12 Regular and uniform hypergraph corresponding to Fig. 4.11.

if there exist constants δ_1 and δ_2 such that $g - \delta_1 \phi(T)$ and $h - \delta_2 \phi(T^T)$ are uniformly decreasing and either one of the following 2 conditions is satisfied:

(1)
$$\alpha \gamma_1 > \sqrt{\theta_1} + \delta_1 \text{ and}$$
$$\frac{1}{2}\sigma_2 \geq 4\beta \frac{\left(\delta_2 - \left(\frac{(\frac{1}{2}\sigma_1 - \alpha)\gamma_2 \beta^2}{4\alpha\sqrt{\theta_1}} + \beta\right)\gamma_2\right)\left(\alpha - \frac{\sqrt{\theta_1} + \delta_1}{\gamma_1}\right)}{\gamma_1 \alpha^2} + \beta$$

(2)
$$\beta \gamma_2 > \sqrt{\theta_2} + \delta_2 \text{ and}$$
$$\frac{1}{2}\sigma_1 \geq 4\alpha \frac{\left(\delta_1 - \left(\frac{(\frac{1}{2}\sigma_2 - \beta)\gamma_1 \alpha^2}{4\beta\sqrt{\theta_2}} + \alpha\right)\gamma_1\right)\left(\beta - \frac{\sqrt{\theta_2} + \delta_2}{\gamma_2}\right)}{\gamma_2 \beta^2} + \alpha$$

where $\theta_1 = \frac{(\frac{1}{2}\sigma_1 - \alpha)\gamma_2^2 \beta^2 (\alpha\gamma_1 - \delta_1)}{4\alpha(\delta_2 - \beta\gamma_2)}$, and $\theta_2 = \frac{(\frac{1}{2}\sigma_2 - \beta)\gamma_1^2 \alpha^2 (\beta\gamma_2 - \delta_2)}{4\beta(\delta_1 - \alpha\gamma_1)}$.

Proof: Consider the case where the first inequality in Theorem 4.5 is an equality, i.e. $\frac{1}{2}\sigma_1 = \frac{4a_2\alpha((a_1 - \alpha)\gamma_1 + \delta_1)}{\gamma_2 \beta^2} + \alpha$. Solving for a_2 we get $a_2 = \frac{(\frac{1}{2}\sigma_1 - \alpha)\gamma_2 \beta^2}{4\alpha((a_1 - \alpha)\gamma_1 + \delta_1)}$. Using this equation for a_2 into the second inequality in Theorem 4.5 we obtain

$$\frac{1}{2}\sigma_2 \geq \frac{4a_1\beta\left(\left(\frac{(\frac{1}{2}\sigma_1 - \alpha)\gamma_2 \beta^2}{4\alpha((a_1 - \alpha)\gamma_1 + \delta_1)} - \beta\right)\gamma_2 + \delta_2\right)}{\gamma_1 \alpha^2} + \beta = f(a_1)$$

$f(a_1 \uparrow \alpha - \frac{\delta_1}{\gamma_1}) = +\infty$, $f(0) = \beta$,

$$f'(a_1) = \frac{4\beta}{\gamma_1 \alpha^2} \left(\delta_2 - \beta\gamma_2 + \frac{\left(\frac{1}{2}\sigma_1 - \alpha\right) \gamma_2^2 \beta^2 (\delta_1 - \alpha\gamma_1)}{4\alpha \left((a_1 - \alpha) \gamma_1 + \delta_1\right)^2} \right)$$

and $f''(a_1) = \frac{2(\frac{1}{2}\sigma_1 - \alpha)\gamma_2^2 \beta^3 (\alpha\gamma_1 - \delta_1)}{((a_1 - \alpha)\gamma_1 + \delta_1)^3 \alpha^3}$. Thus for $0 < a_1 < \alpha - \frac{\delta_1}{\gamma_1}$, $f''(a_1) > 0$ and $f'(a_1) = 0$ at $a_1 = \alpha - \frac{\delta_1 + \sqrt{\theta_1}}{\gamma_1}$ which is a minimum of f. To ensure that $a_1 > 0$, we need $\alpha\gamma_1 - \delta_1 > \sqrt{\theta_1}$. The second inequality in Theorem 4.5 then becomes the second inequality in Condition (1). The second case where the second inequality in Theorem 4.5 is an equality follows by substituting α, δ_1, γ_1, σ_1 with β, δ_2, γ_2, σ_2 respectively. □

In the coupled array in Example 4.2, the coupling elements are passive circuits and thus one of the δ's can be chosen to be zero, and for a class of regular and uniform hypergraphs we have the following result.

Theorem 4.7 *Consider an array of Chua's oscillators similarly coupled as Fig. 4.11 with an underlying connected hypergraph which is regular and uniform. Assume the same circuit parameters as in Example 4.2. If $(\beta - \frac{1}{2}\sigma_2)\alpha^3 = (\alpha - \frac{1}{2}\sigma_1)\beta^3$, then the array synchronizes if*

$$0 < R_1 < \frac{1}{G_N} \left(\alpha + \frac{\beta \left(\frac{1}{2}\sigma_1 - \alpha\right)}{4\alpha} \left(2 + \frac{\gamma_1}{\gamma_2} + \frac{\gamma_2}{\gamma_1} \right) \right)$$

Proof: The case of $\sigma_1 = 2\alpha$ and $\sigma_2 = 2\beta$ is covered by Corollary 4.1. Let us assume that $\sigma_1 < 2\alpha$ which by hypothesis implies that $\sigma_2 < 2\beta$. Since δ_2 can be arbitrarily small, let us first choose a_1 and a_2 such that $\frac{1}{2}\sigma_2 = \frac{4a_1 \beta \gamma_2 (a_2 - \beta)}{\gamma_1 \alpha^2} + \beta$. This implies that $a_1 = \frac{(\frac{1}{2}\sigma_2 - \beta)\gamma_1 \alpha^2}{4\beta\gamma_2(a_2 - \beta)}$. Since $a_1 > 0$, this implies that $a_2 < \beta$. It is clear that a_1 and a_2 can be perturbed arbitrarily small to satisfy $\frac{1}{2}\sigma_2 \geq \frac{4a_1\beta((a_2-\beta)\gamma_2+\delta_2)}{\gamma_1\alpha^2} + \beta$ for sufficiently small δ_2.

The largest δ_1 which satisfies the conditions of Theorem 4.5 is obtained by making the inequality on σ_1 an equality which (after substituting the above equation for a_1) is equal to:

$$\delta_1 = \frac{(\frac{1}{2}\sigma_1 - \alpha)\gamma_2\beta^2}{4a_2\alpha} - \frac{(\frac{1}{2}\sigma_2 - \beta)\gamma_1^2\alpha^2}{4\beta\gamma_2(a_2 - \beta)} + \alpha\gamma_1$$

By taking the derivative with respect to a_2 we see that the value of δ_1 is maximized when

$$\frac{-(\frac{1}{2}\sigma_1 - \alpha)\gamma_2\beta^2}{4\alpha a_2^2} + \frac{(\frac{1}{2}\sigma_2 - \beta)\gamma_1^2\alpha^2}{4\gamma_2\beta(a_2 - \beta)^2} = 0$$

Since $(\beta - \frac{1}{2}\sigma_2)\alpha^3 = (\alpha - \frac{1}{2}\sigma_1)\beta^3$, this implies that $\gamma_1^2 a_2^2 = \gamma_2^2 (a_2 - \beta)^2$, which has a solution $a_2 = \frac{\gamma_2\beta}{\gamma_1+\gamma_2} < \beta$. This corresponds to

$$\frac{\delta_1}{\gamma_1} = \alpha + \frac{\beta(\frac{1}{2}\sigma_1 - \alpha)}{4\alpha}\left(2 + \frac{\gamma_1}{\gamma_2} + \frac{\gamma_2}{\gamma_1}\right)$$

so the array synchronizes if $\frac{\delta_1}{\gamma_1} < \alpha + \frac{\beta(\frac{1}{2}\sigma_1-\alpha)}{4\alpha}\left(2 + \frac{\gamma_1}{\gamma_2} + \frac{\gamma_2}{\gamma_1}\right)$. Since $\frac{\delta_1}{\gamma_1} = R_1 G_N$, the theorem is proved. □

Corollary 4.2 *Consider an array of Chua's oscillators similarly coupled as Fig. 4.11 with an underlying hypergraph which is self-dual and regular with vertex degree α and algebraic connectivity $\sigma > 0$. Assume the same circuit parameters as in Example 4.2. The array synchronizes if $0 < R_1 < \frac{1}{G_N}\left(\alpha + \frac{(\frac{1}{2}\sigma - \alpha)}{4}\left(2 + \frac{C}{C_1} + \frac{C_1}{C}\right)\right)$.*

For example, by choosing optimal a_1 and a_2 according to Corollary 4.2, the array in Example 4.2 synchronizes if $0 < R_1 < \frac{1}{G_N}\left(3 - \frac{2+\frac{C}{C_1}+\frac{C_1}{C}}{12}\right)$.

Example 4.3 Consider the array of coupled Chua's oscillators where the underlying regular and uniform hypergraph is such that each of the n vertices is connected to $n - 1$ edges out of a total of n edges (Fig. 4.13). The algebraic connectivity of both the hypergraph and its dual is $2\left(n - 1 - \frac{1}{n-1}\right)$ and by Corollary 4.2 the array synchronizes if $0 < R_1 < \frac{1}{G_N}\left(n - 1 - \frac{2+\frac{C}{C_1}+\frac{C_1}{C}}{4(n-1)}\right)$.

Example 4.4 Consider an array of coupled Chua's oscillators which is fully coupled, i.e. each cell (Chua's oscillator) is coupled to every coupling element and vice versa. Let the number of cells and coupling elements be n and m respectively. The circuit equations are the same as before, except that the incidence matrix is an n by m matrix of 1's. The corresponding hypergraph is shown in Fig. 4.14 for $n = 4$, $m = 5$. The algebraic connectivities of the hypergraph and its dual are $2m$ and $2n$ respectively. Applying Theorem 4.7 (or Corollary 4.1) the array synchronizes if $0 < R_1 < \frac{m}{G_N}$. This

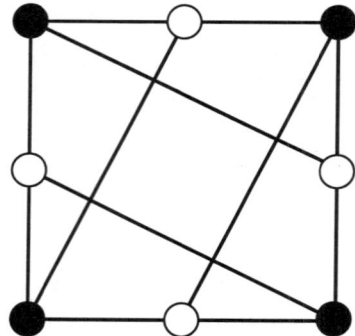

Fig. 4.13 Almost complete hypergraph of 4 vertices and 4 edges.

result could also be obtained by applying the circuit theoretical techniques discussed in Section 4.1.

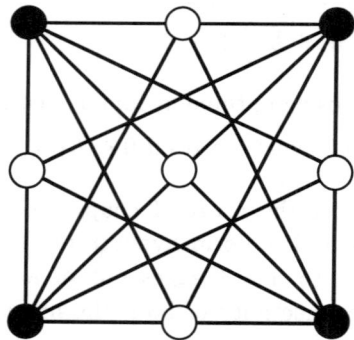

Fig. 4.14 Complete hypergraph of 4 vertices and 5 edges.

4.3 Two identical systems coupled by dynamic coupling

Consider the special case of two identical systems coupled by dynamic coupling with state equations given by:

$$\begin{aligned}\dot{x}_1 &= g(x_1, y)\\ \dot{x}_2 &= g(x_2, y)\\ \dot{y} &= \hat{h}(y, x_1, x_2)\end{aligned} \qquad (4.10)$$

By Theorem 2.1 $x_1 \to x_2$ if $\dot{x} = g(x, u)$ is asymptotically stable for all u. Consider the unidirectional coupling case where $\dot{y} = \hat{h}(y, x_1)$. The resulting system is then equivalent to the Pecora-Carroll subsystem decomposition (Section 2.1).

Chapter 5
Graph Topology and Synchronization

Definition 5.1 The *synchronization subspace* corresponding to Eq. (3.2) is the linear subspace of vectors of the form $(x, ..., x)$.

From Definition 3.1, the trajectory of a synchronizing coupled system (3.2) converges towards a (not necessarily unique) solution in the synchronization subspace of the form $x_s(t) = (x(t), \ldots, x(t))$ as $t \to \infty$. Some of these $x_s(t)$ are *not* necessarily a solution of Eq. (3.2). However, in the static linear coupling case (3.8) with G a zero row sum matrix, a $x_s(t)$ exists which is also a solution of Eq. (3.8). Furthermore, $x_s(t) = (x(t), \ldots, x(t))$ is given by $x(t)$ being a trajectory of the uncoupled individual system $\dot{x}_i = f(x_i, t)$. This is true since G has zero row sums and thus the synchronization subspace is in the nullspace of the linear operator $G \otimes D$, i.e. the coupling term $(G \otimes D)x$ is zero for any solution in the synchronization subspace. When G has constant row sums μ for all rows, a simple substitution $f(x,t) \to f(x,t) - \mu Dx$ results in a matrix G with zero row sums.

In the case of dynamic coupling, because of the dynamics of the coupling systems, the cell systems' trajectories do not necessary converge towards trajectories of uncoupled cell systems.

Corollaries 3.3, 3.2 and 3.4 and Theorem 3.15 relate synchronization criteria to the nonzero eigenvalues of a coupling matrix G. Consider the case where this matrix is proportional to the Laplacian matrix of the underlying graph, i.e. $G = \epsilon L$ where L is the Laplacian matrix of the underlying graph and $\epsilon > 0$ is the *coupling strength*. These synchronization criteria do not depend on the underlying graph specifically, but only on the spectra of its Laplacian matrix. On the other hand, the relationship between the eigenvalues of Laplacian matrices and properties of the corresponding graphs has been extensively studied [81, 82]. In this chapter we explore some aspects of this relationship and their implications for synchronizing coupled arrays of circuits and systems.

5.1 Some coupling configurations

Some common coupling configurations for array of coupled systems studied in the literature are:

(1) Complete graph K_n (Fig. 5.1). The Laplacian matrix is given by

$$\begin{pmatrix} n-1 & -1 & -1 & \cdots & & -1 \\ -1 & n-1 & -1 & \cdots & & -1 \\ \vdots & & \ddots & & & \vdots \\ -1 & -1 & \cdots & n-1 & -1 \\ -1 & -1 & \cdots & -1 & n-1 \end{pmatrix}$$

with eigenvalues 0 and n (with multiplicity $n-1$).

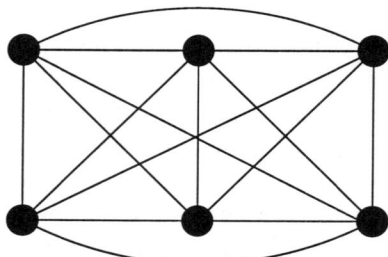

Fig. 5.1 Complete graph K_n where every vertex is connected to every other vertex (for the case $n = 6$).

(2) Path graph P_n (Fig. 5.2). The Laplacian matrix is given by

$$\begin{pmatrix} 1 & -1 & & & & \\ -1 & 2 & -1 & & & \\ & \ddots & \ddots & \ddots & & \\ & & & -1 & 2 & -1 \\ & & & & -1 & 1 \end{pmatrix}$$

with eigenvalues [13] $2 + 2\cos\left(\frac{i\pi}{n}\right), i = 1, \ldots, n$.

Fig. 5.2 Path graph P_n (for the case $n = 6$).

(3) Directed path graph DP_n (Fig. 5.3). The corresponding connectivity matrix is given by

$$\begin{pmatrix} 0 & & & & \\ -1 & 1 & & & \\ & -1 & 1 & & \\ & & \ddots & \ddots & \\ & & & -1 & 1 \end{pmatrix}$$

with eigenvalues 1 (multiplicity $n - 1$) and 0.

Fig. 5.3 Directed path graph DP_n (for the case $n = 6$).

(4) Cycle graph C_n (Fig. 5.4). The Laplacian matrix is circulant and is given by

$$\begin{pmatrix} 2 & -1 & & & & -1 \\ -1 & 2 & -1 & & & \\ & \ddots & \ddots & \ddots & & \\ & & -1 & 2 & -1 \\ -1 & & & & -1 & 2 \end{pmatrix}$$

with eigenvalues $4\sin^2\left(\frac{i\pi}{n}\right), i = 1 \ldots, n$.

Fig. 5.4 Cycle graph C_n (for the case $n = 6$).

(5) Directed cycle graph DC_n (Fig. 5.5). The corresponding connec-

tivity matrix is circulant and is given by

$$\begin{pmatrix} 1 & & & & & -1 \\ -1 & 1 & & & & \\ & -1 & 1 & & & \\ & & \ddots & \ddots & & \\ & & & & -1 & 1 \end{pmatrix}$$

with eigenvalues $1 - \cos\left(\frac{2j\pi}{n}\right) - i\sin\left(\frac{2j\pi}{n}\right), j = 0\ldots, n-1$ where $i = \sqrt{-1}$.

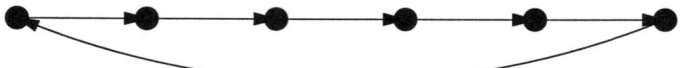

Fig. 5.5 Directed cycle graph DC_n (for the case $n = 6$).

(6) Complete bipartite graph $K_{m,n}$ where the two parts consists of m and n vertices (Fig. 5.6). The Laplacian matrix is given by

$$\begin{pmatrix} m & & & -1 & \ldots & -1 \\ & \ddots & & \vdots & \vdots & \vdots \\ & & m & -1 & \ldots & -1 \\ -1 & \ldots & -1 & n & & \\ \vdots & \vdots & \vdots & & \ddots & \\ -1 & \ldots & -1 & & & n \end{pmatrix}$$

with eigenvalues 0, m (with multiplicity $n-1$), n (with multiplicity $m-1$, and $n+m$.

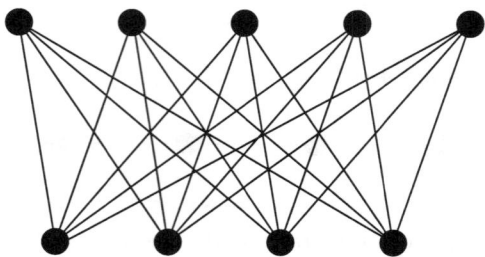

Fig. 5.6 Complete bipartite graph $K_{m,n}$ (for the case $n = 5, m = 4$).

(7) Star graph S_n (Fig. 5.7). The Laplacian matrix is given by

$$\begin{pmatrix} n-1 & -1 & \cdots & -1 \\ -1 & 1 & & \\ \vdots & & \ddots & \\ -1 & & & 1 \end{pmatrix}$$

with eigenvalues 0, 1 (with multiplicity $n-2$), and n.

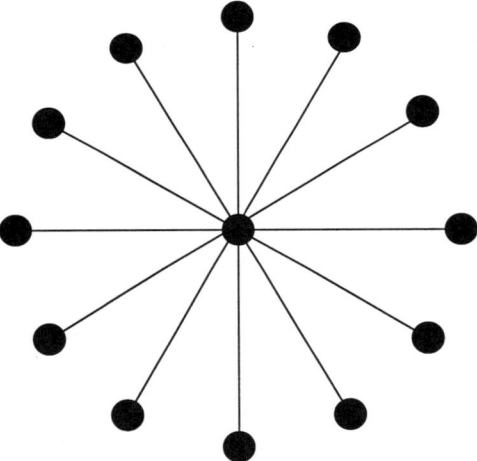

Fig. 5.7 Star graph S_n (for the case $n = 13$).

In this chapter we are interested in how the eigenvalues of such coupling configurations behave as $n \to \infty$ and their relation to synchronization criteria.

5.2   Continuous time systems

In order to apply Corollaries 3.3 and 3.4 the coupling matrix G needs to be an irreducible and symmetric matrix with zero row sums and nonpositive off-diagonal elements, a condition which is satisfied if $G = \epsilon L$, $\epsilon > 0$, and L is the Laplacian matrix of a connected graph.

All nonzero eigenvalues of $G = \epsilon L$ are larger than $\epsilon \alpha$, where α is the least nonzero eigenvalue of L. If the other conditions in Corollary 3.4 are satisfied, then the coupled array synchronizes for large enough ϵ and that the required coupling strength ϵ required is inversely proportional to the least nonzero eigenvalue of L, or the algebraic connectivity of the graph. In this case, lower bounds on the algebraic connectivity of a graph in terms of graph properties are useful in deriving synchronization criteria [83]:

Lemma 5.1   *The algebraic connectivity $a(\mathcal{G})$ of a graph $\mathcal{G}$ satisfies:*

(1) $a(\mathcal{G}) \geq 2\Delta_{\min} - n + 2$

(2) $a(\mathcal{G}) \geq 2e_c \left(1 - \cos\left(\frac{\pi}{n}\right)\right)$

(3) $a(\mathcal{G}) \geq \Delta_{\max} - \sqrt{\Delta_{\max}^2 - i_s^2}$

(4) $a(\mathcal{G}) \geq \frac{4}{n\mathrm{diam}}$

(5) $a(\mathcal{G}) \geq \frac{2}{(n-1)\bar{\rho} - \frac{n-2}{2}}$

where $\Delta_{\min}$ and $\Delta_{\max}$ are the smallest and largest vertex degrees, e_c is the edge connectivity, i_s is the isoperimetric number, diam is the diameter, $\bar{\rho}$ is the mean distance, and n is the number of vertices.

Proof: These bounds follow from the results in [84, 85, 86]. □

Next we consider some upper bounds for $a(\mathcal{G})$. In Fig. 3.1 Chua's oscillators are coupled via linear resistors. Suppose that we have a limited number of coupling resistors available such that each oscillator can be coupled to at most k other oscillators. This corresponds to a graph where each vertex is connected to at most k other vertices. A larger algebraic connectivity implies that a smaller conductance is needed for the coupling resistors in order to synchronize the array. Since the algebraic connectivity of a graph can only increase by adding more edges (Lemma A.7), the largest algebraic connectivity is obtained when each vertex is coupled to k other vertices*.

For a fixed k, what can we say about the algebraic connectivity of graphs whose maximal vertex degree is k, as the number of vertices n goes to infinity? From Lemma A.4, the algebraic connectivity is less than or equal to $\frac{nk}{n-1}$ ($\approx k$ for large n). For $k = 2$, the path and cycle graphs in

---

*This is assuming that either k or the number of vertices is even so that a k-regular graph exists.

Section 5.1 shows that the algebraic connectivity decreases to 0 as $O\left(\frac{1}{n^2}\right)$. The path graph and its higher dimensional analog, the grid graph, are obtained by discretizing the continuous Laplacian operator. A grid graph is defined as a graph where the vertices lies on an d-dimensional integer lattice and each vertex is connected to its nearest neighbors. It turns out that for the grid graph, the algebraic connectivity decreases to 0 as $n \to \infty$. The diameter c of a grid graph as a function of the number of vertices n is on the order of $\Omega(\sqrt[d]{n})$. The maximum vertex degree is $\Delta = 2d$. Mohar [86] shows that the algebraic connectivity is less than

$$\frac{\Delta \ln(n-1)}{2(c-2) - \ln(n-1)}$$

if $2(c-2) - \ln(n-1) > 0$ and thus the algebraic connectivity goes to zero as $n \to \infty$ for the grid graph.

The same argument can be used when the vertices are connected to neighbors in an arbitrary local neighborhood. Thus graphs where the vertices are arranged on this lattice and each vertex is coupled to vertices in a local neighborhood will have the algebraic connectivity decreasing to zero for $n \to \infty$.

This might lead one to conjecture that the algebraic connectivity approaches zero as the number of vertices goes to infinity if the edge density[†] is bounded.

This conjecture turns out to be false. From the above discussion, to generate graphs with bounded edge density where the algebraic connectivity is bounded away from zero as the number of vertices increases, the diameter of the graph cannot increase faster than $O(\ln(n))$. This implies that the graph cannot only have local connections. It is intuitively clear that one type of graphs where this occurs are random graphs. We give here several results which confirm this for large k. In fact, the following results show that the algebraic connectivity grows as $O(k)$ for k-regular random graphs as $n \to \infty$ for large k, even as $\frac{k}{n} \to 0$.

Friedman et al. [77] proved the following theorem regarding random regular graphs:

Theorem 5.1 *Let $\mathcal{G}$ be a random 2d-regular graph with n vertices. Let $0 = \lambda_1 \leq \lambda_2 \leq \ldots \leq \lambda_n$ be the eigenvalues of its Laplacian matrix. Then*

[†]The edge density is defined as the number of edges divided by the number of vertices. The edge density of a k-regular graph is $\frac{k}{2}$.

for any $\beta > 1$,

$$\rho \geq \left(2\sqrt{2d-1} + 2\log d + O(1) + O\left(\frac{d^{3/2}\log\log n}{\log n}\right)\right)\beta \quad (5.1)$$

with probability less than or equal to

$$\frac{\beta^2}{n^{2\lfloor\sqrt{2d-1}/2\rfloor\log\beta/\log d}}$$

where $\rho = \max(|2d - \lambda_2|, |2d - \lambda_n|)$.

We can draw the following conclusions from Theorem 5.1:

(1) For a $2d$-regular random graph with large d and n, $\lambda_2 > 2d - 2\sqrt{2d-1} - 2\log d$ with high probability.
(2) For a $2d$-regular random graph with large d and n, $\lambda_n < 2d + 2\sqrt{2d-1} + 2\log d$ with high probability.
(3) For a $2d$-regular random graph with large d and n, $\frac{\lambda_n}{\lambda_2} \leq 1 + \frac{2\sqrt{2d-1}+2\log d}{d-\sqrt{2d-1}-\log d}$ with high probability.

This implies that that for random k-regular graphs where k is even and k, n large, the algebraic connectivity λ_2 is larger than $k - O(\sqrt{k})$[‡], λ_n is less than $k + O(\sqrt{k})$ and $1 \leq \frac{\lambda_n}{\lambda_2} \leq 1 + O\left(\frac{1}{\sqrt{k}}\right)$ with high probability.

By combining the results of Mohar [85] and Bollobás [88], it can also be shown that the algebraic connectivity of almost every k-regular random graph is bounded away from zero as $n \to \infty$ provided k is large enough.

Theorem 5.2 *For any $\epsilon > 0$, the algebraic connectivity of almost every k-regular random graph is larger than $\left(1 - \frac{\sqrt{3}}{2} - \epsilon\right)k$ as $n \to \infty$ for a large enough k.*

Proof: According to [88] the isoperimetric number of almost every k-regular random graph satisfies $i \geq \frac{k}{2} - \sqrt{k\ln 2}$ for large enough k. According to [85] the algebraic connectivity α of a graph satisfies $i \leq \sqrt{\alpha(2k-\alpha)}$. Combining these two inequalities we find that

$$\alpha \geq k - \sqrt{\frac{3}{4}k^2 - k\left(\ln 2 - \sqrt{k\ln 2}\right)}.$$

□

[‡]This conclusion was also obtained by Gade [87].

Other examples of graphs where this behavior is observed are the small world networks considered in [89] where random connections are incrementally added to an initial regular lattice with only local connections.

5.3 Discrete-time systems

Consider the array in Eq. (3.18) where D is a positive multiple of the identity matrix. As discussed in Section 3.4, Corollary 3.9 implies that ϵ exists which synchronizes the array provided the ratio $\frac{\lambda_n}{\lambda_2}$ between the largest and the smallest nonzero eigenvalues of the Laplacian matrix L is close enough to 1. Since L is symmetric, $\lambda_n = \max_{x \neq 0} \frac{x^T L x}{x^T x}$ by the Courant-Fischer theorem [90]. Suppose the maximal vertex degree is k. This means that $L_{i,i} = k$ for some i. Setting $x = e_i$ in the above equation for λ_n we see that $\lambda_n \geq k$. In fact, in [91] it was shown that $\lambda_n \geq k+1$ if the graph has at least one edge.

For graphs where each vertex is connected to vertices in a local neighborhood $\lambda_2 \to 0$ as $n \to \infty$ (Section 5.2). Therefore the ratio $\frac{\lambda_n}{\lambda_2} \to \infty$ as $n \to \infty$.

On the other hand the ratio $\frac{\lambda_n}{\lambda_2}$ approaches 1 as $n \to \infty$ for random k-regular graphs (Section 5.2). This means that ϵ exists which synchronizes a randomly coupled array. This is even true when the ratio $\frac{k}{n}$ is arbitrarily small. Since a small $\frac{k}{n}$ indicates a sparsely coupled array, this shows that arbitrarily sparsely coupled array of discrete-time systems can be made synchronizing when the cells are coupled randomly and the array is large enough. This is in contrast to the continuous-time case (Eq. (3.8)) which by Corollary 3.4 can always be made synchronizing by choosing the appropriate matrix D if f has bounded Jacobian matrices and the underlying graph is connected.

Therefore for both the continuous time systems and discrete time systems, the sufficient conditions for synchronization are easier to satisfy if the vertices are coupled not only to vertices in local neighborhoods, but also to vertices far away, as is the case in random coupling.

Theorem 5.1 says that the nonzero eigenvalues of random k-regular graphs lies in the range $[k - 2\sqrt{k}, k + 2\sqrt{k}]$ with high probability. In [76] random directed graphs are considered in the context of coupled map lattices. The corresponding matrix $V - A$ where V is the row sum of the adjacency matrix A is not necessarily symmetric, i.e. the coupling is not necessarily reciprocal, but this class of matrices includes the Laplacian matrices of random graphs as a subset. It was shown in [87] that the nonzero eigenvalues of $V - A$ lie in the ball of radius $\sqrt{k}$ centered at k with high probability.

For dynamic coupling, the underlying coupling topology is best expressed as a hypergraph. There are coupling topologies for static coupling which is also best expressed as a hypergraph [70]. With Δ-Y transformations [92] and its generalizations, this case can be reduced to a weighted graph, but sometimes it is still instructive to study the properties of the underlying hypergraph.

5.4 Graph coloring via synchronized array of coupled oscillators

For 2-colorable graphs, a simple algorithm suffices to generate the 2-coloring. First assign an arbitrary vertex one color, then assign its neighbors the second color, then assign its neighbors' neighbors the first color, etc. until the entire graph is colored[§]. This algorithm can also be used to check whether the graph is 2-colorable or not, as the resulting coloring is not a valid coloring of the graph if the graph is not 2-colorable, i.e. at some step of the algorithm, a vertex has the same color as one of its (previously colored) neighbors. For graphs where the chromatic number is higher than 2, things are much more complicated. In fact, the problem of coloring a graph with the minimal number of colors is NP-hard.

In this section, we show how synchronized array of circuits can be used to find a good coloring of graphs. The coupling topology of an array of circuits can be thought of as graphs with the circuits representing vertices. Recall from Chapter 3 that the adjacency matrix A of the underlying connectivity graph of Eq. (3.8) is given by

$$A_{ij} = \begin{cases} 1 & i \neq j, G_{ij} \neq 0 \\ 0 & \text{otherwise} \end{cases}$$

[§]For disconnected graphs, this procedure has to be repeated for each connected component.

5.4.1 Coloring two-colorable graphs

The following theorem [93] gives conditions when the coupled array synchronizes into two clusters with opposite phases when the connectivity graph is connected and bipartite. By assigning two colors to the two phases we obtain a two coloring of the underlying graph.

Theorem 5.3 *Suppose that the following conditions are satisfied in Eq. (3.8):*

(1) $G = -\alpha(A + diag(v(i)))$, *where $\alpha > 0$ and A is the adjacency matrix of the underlying graph and $diag(v(i))$ is the diagonal matrix with the degrees of the vertices on the diagonal,*
(2) $f(x,t) = -f(-x,t)$ *for all x, t (f is odd-symmetric with respect to x),*
(3) *there exists matrices D and V such that $VD \geq 0$ and $f(x,t) - Dx$ is V-uniformly decreasing.*
(4) *The underlying graph is connected and 2-colorable.*
(5) *The algebraic connectivity of the underlying graph is larger than or equal to $\frac{1}{\alpha}$.*

Then if x_i and x_j are adjacent in the underlying graph, they synchronize out of phase, i.e. $x_i \to -x_j$ as $t \to \infty$. In other words, the phases of the oscillators generate a 2-coloring of the graph.

Proof: For a 2-coloring of the graph, let B be the diagonal matrix such that B_{ii} is equal to 1 or -1 depending on the color of the i-th vertex. Since $A_{ij} \neq 0$ implies $B_{ii} = -B_{jj}$, we have $(BAB)_{ij} = B_{ii}A_{ij}B_{jj} = -A_{ij}$. Therefore $BAB + Bdiag(v(i))B = -A + diag(v(i))$. Using the state transformation $x \to (B \otimes I)x$ and the fact that $f(-x,t) = -f(x,t)$, the state equation (3.8) can be rewritten as:

$$\dot{x} = \begin{pmatrix} f(x_1, t) \\ \vdots \\ f(x_m, t) \end{pmatrix} + (B \otimes I)(G \otimes D)(B \otimes I)x \qquad (5.2)$$

The matrix $(B \otimes I)(G \otimes D)(B \otimes I)$ is equal to $BGB \otimes D = (-\alpha(diag(v(i)) - A)) \otimes D$. The algebraic connectivity condition implies that all the nonzero eigenvalues of $-\alpha(diag(v(i))) - A$ are less than or equal to -1 and thus the conclusion follows from Corollary 3.4. □

Since for a graph of N vertices the algebraic connectivity is larger than or equal to $2\left(1-\cos\left(\frac{\pi}{N}\right)\right)$ by Lemma 5.1, we can choose $\alpha > \frac{1}{2(1-\cos(\frac{\pi}{N}))}$ to ensure that condition (5) in Theorem 5.3 is satisfied.

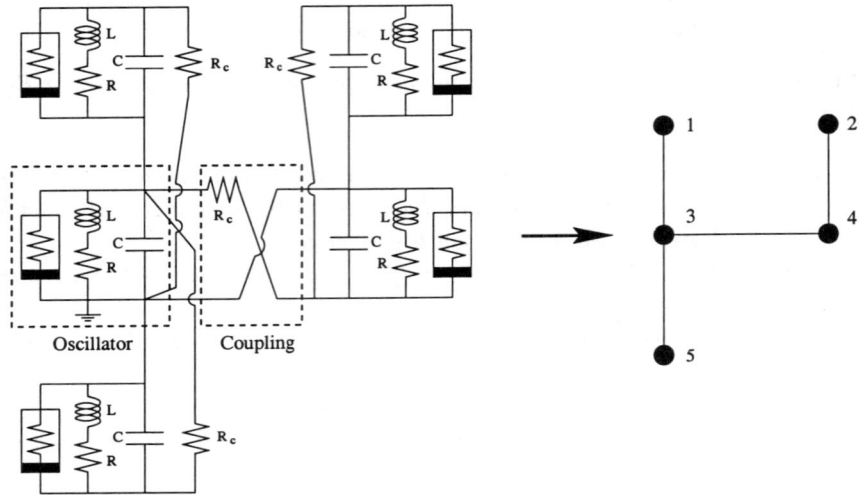

Fig. 5.8 Arrays of coupled relaxation oscillators and its associated graph.

Consider an array of coupled relaxation oscillators and its associated graph G shown in Fig. 5.8. The state equations for this coupled system are given by:

$$\begin{aligned} C\tfrac{dv_j}{dt} &= -i_j - g(v_j) - \tfrac{1}{R_c}(Gv)_j \\ L\tfrac{di_j}{dt} &= v_j - \tfrac{i_j}{R} \end{aligned} \quad (5.3)$$

where v_j is the voltage across the capacitors, i_j is the current through the inductors, $v = (v_1, v_2, \ldots, v_N)^T$, $N = 5$ and $g(v) = -\epsilon(v - \tfrac{v^3}{3})$ is the nonlinear v-i characteristic of the nonlinear resistor. For the coupling configuration in Fig. 5.8, the matrix G is given by:

$$G = \begin{pmatrix} 1 & 0 & 1 & 0 & 0 \\ 0 & 1 & 0 & 1 & 0 \\ 1 & 0 & 3 & 1 & 1 \\ 0 & 1 & 1 & 2 & 0 \\ 0 & 0 & 1 & 0 & 1 \end{pmatrix}$$

We can apply Theorem 5.3 to the coupled oscillators in Eq. (5.3) when we choose

$$D = \begin{pmatrix} \frac{\epsilon+\delta}{C} & 0 \\ 0 & 0 \end{pmatrix}, \quad V = \begin{pmatrix} C & 0 \\ 0 & L \end{pmatrix} \quad R_c = \frac{1}{(\epsilon+\delta)\alpha}, \quad C, L > 0$$

$f(x,t) - Dx$ is V-uniformly decreasing for any $\delta > 0$ (Section 2.3.3). It then follows that the system in Eq. (5.3) synchronizes and generates a 2-coloring if $R_c < \frac{a}{\epsilon}$ and the graph is connected and 2-colorable. The number a is the algebraic connectivity of the graph which in the case of Fig. 5.8 is equal to 0.5858. Since the graph in Fig. 5.8 is 2-colorable and connected, the system in Fig. 5.8 synchronizes and generates a 2-coloring if $R_c < \frac{0.5858}{\epsilon}$.

Note that the coupling in Fig. 5.8 is a resistive 2-port where the leads are "twisted" in the output port. A circuit theoretical interpretation of Theorem 5.3 for Fig. 5.8 is that if certain oscillators can be turned upside down as to "untwist" the coupling, then we would result in an array of oscillators with diffusive coupling and Corollary 3.4 can be applied directly. The oscillator can be turned upside down and its dynamics remain the same since all the components are bilateral¶. For 2-colorable graphs, this untwisting can be done exactly when the oscillators of one color are turned upside down.

5.4.2 Coloring arbitrary graphs

In this section we present an array of coupled oscillators for coloring arbitrary graphs. For graphs with chromatic numbers greater than 2, simulation results with small graphs indicate that the array in Eq. (3.8) with

$$f\left(\begin{pmatrix} y \\ z \end{pmatrix}, t\right) = \begin{pmatrix} -z + 5\left(y - \frac{y^3}{3}\right) \\ y \end{pmatrix}, \quad D = \begin{pmatrix} 0 & 0 \\ 0 & 1 \end{pmatrix} \quad (5.4)$$

$$G = -0.21 \text{diag}(\tfrac{1}{v(i)})(A + \text{diag}(v(i)))$$

colors the graph with the minimal number of colors most of the time. Here $\text{diag}(\frac{1}{v(i)})$ and $\text{diag}(v(i))$ denote diagonal matrices with $\frac{1}{v(i)}$ and $v(i)$ on the diagonal respectively.

As synchronization between all the oscillators can only occur if the graph is connected, we will only consider connected graphs. The initial conditions

---

¶This results in the vector field being symmetric with respect to the origin.

are chosen at random from the interval $[-1, 1]$. This system can be implemented by substituting the oscillator shown in Fig. 5.9 for the oscillators in Fig. 5.8 where $\alpha_{ii} = \frac{0.21}{v(i)}$. For regular graphs, $\text{diag}(1/v(i))$ is a multiple of the identity matrix, and by choosing $R_c = \frac{v(i)}{0.21}$ we can eliminate the controlled current source in Fig. 5.9.

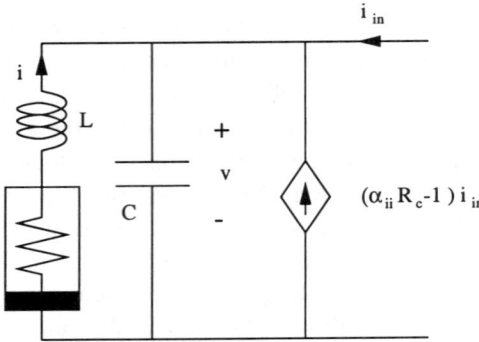

Fig. 5.9 Substituting this oscillator for the oscillators in Fig. 5.8 results in dimensionless state equations of the form Eq. (3.8) and Eq. (5.4). The parameter α_{ii} is equal to $\frac{0.21}{v(i)}$.

Although Theorem 5.3 cannot be applied to this case when G is 2-colorable, computer simulations indicate that the array synchronizes to the proper 2-coloring. For more general graphs, a more complicated algorithm is needed to map a synchronized state of the oscillators onto a coloring scheme. Let $0 \leq \phi(i) < 2\pi$ be the phase of the i-th oscillator at the synchronized state. For each $0 \leq d \leq 2\pi$, associate a mapping which maps ϕ onto a coloring c of the graph G. We denote this map as the decoding map μ_d. The map μ_d is defined such that vertices v_i and v_j are colored the same color if and only if the difference between the phase of the i-th oscillator and the j-th oscillator is less than or equal to d and v_i and v_j are not adjacent in the graph. The phase difference is computed modulo 2π. More precisely

$$\mu_d(v_i) = \mu_d(v_j) \Leftrightarrow (h(\phi(i), \phi(j)) \leq d \quad \text{AND} \quad v_i \text{ is not adjacent to } v_j)$$

where the function h calculates the difference between the phases of oscillators i and j (modulo 2π) and is defined as $h(a, b) = \min(|a - b|, |a - b + 2\pi|, |a - b - 2\pi|)$. This in general does not uniquely determine μ_d even up to a permutation of the colors. Note that μ_0 is the map used to determine the 2-coloring in Theorem 5.3.

It's clear that each μ_d generates a valid coloring of the graph. For graphs which are not 2-colorable, most of the time at the synchronized state all the oscillators will have different phases, so that μ_0 will generate the trivial coloring of assigning each vertex a different color. In our graph coloring algorithm, for each synchronized state we increase d from 0 or decrease d from 2π, until a valid coloring is found with the minimal number of colors. The pseudocode for this algorithm is shown in Fig. 5.10. Thus the steady state form a relaxed coloring of the graph, where we assign to each vertex the phase of the corresponding oscillator. What is desired is that adjacent vertices have phases far apart from each other. The property of being "far apart" is captured by $h(\phi(i), \phi(j)) > d$. The algorithm tries to find d which gives a minimal valid coloring. An alternative algorithm can be obtained by replacing Simulate system for T seconds with Simulate system until steady state and removing the WHILE and ENDWHILE statements.

The constant δ should be taken to be inversely proportional to the number of vertices. We have simulated Eq. (4.4) with the oscillators defined as Eq. (5.4) with G corresponding to various small graphs with chromatic number 3. Most of the time, the system generates a 3-coloring. In particular, we have conducted experiments with 300 graphs with the number of vertices ranging from 4 to 16. The graphs are generated randomly using the algorithm in [94] where the edge density p is a uniform random variable in the range $[2/N, 1]$. The generated graphs have either chromatic number 2 or 3. The maximum time simulated (maxtime in Fig. 5.10) is 2000, which corresponds to approximately 160 cycles of oscillations. The system colored all 2-colorable graphs (except for one) correctly, and of the remaining graphs which have chromatic number 3, it colored more than 79% with three colors, and more than 94% with less than or equal to 6 colors. Of the graphs which were minimally colored with 3 colors, the correct coloring was found in less than 100 cycles.

As an example, consider the 3-colorable graph with 12 vertices as shown in Fig. 5.11. The waveforms of x_i are shown in Fig. 5.12. The waveforms are translated in the vertical axis to avoid clutter, with oscillator 1 at the bottom and oscillator 12 at the top. At $t = 0$, the random initial conditions result in random phases for the 12 oscillators. Near the end of the simulation ($t = 100$), we can clearly see that the phases can be grouped into 3 groups, with oscillators 1-4 in one group, oscillators 5-8 in a second group, and

```
BEGIN
  totaltime = 0
  WHILE totaltime < maxtime DO
    Simulate system for T seconds.
    Get state vector at end of simulation and calculate φ(i).
    d = 0
    DO
       d = d + δ
       Use μ_d to generate coloring.
       IF coloring generated by μ_d is satisfactory
          STOP and RETURN coloring generated by μ_d.
       ENDIF
    UNTIL d = 2π
    totaltime = totaltime + T
  ENDWHILE
  RETURN coloring generated by μ_d.
END
```

Fig. 5.10 Pseudocode of algorithm to generate coloring from phase information of the oscillators. An alternative algorithm can be obtained by replacing "Simulate for T seconds" with "Simulate until steady state" and removing the WHILE and ENDWHILE statements.

oscillators 9-12 in the third group. Assigning oscillators in each group to one color results in a valid 3-coloring of the graph.

As a second example, consider the connected bipartite graph with 16 vertices shown in Fig. 5.13. The waveforms of x_i are shown in Fig. 5.14. It is clear that after a few cycles the first 8 oscillators are synchronized in phase, while the other 8 oscillators are synchronized at opposite phase, creating a valid 2-coloring for the graph.

5.4.3 Antivoter model for graph coloring

The antivoter model proposed in [95] for graph coloring shares some similarities and differences with the present approach of coupled oscillators. In the antivoter model, adjacent vertices repel each other at random intervals. We believe that in the system of coupled oscillators, adjacent oscillators also "repel" each other so that they will not share the same phase. How-

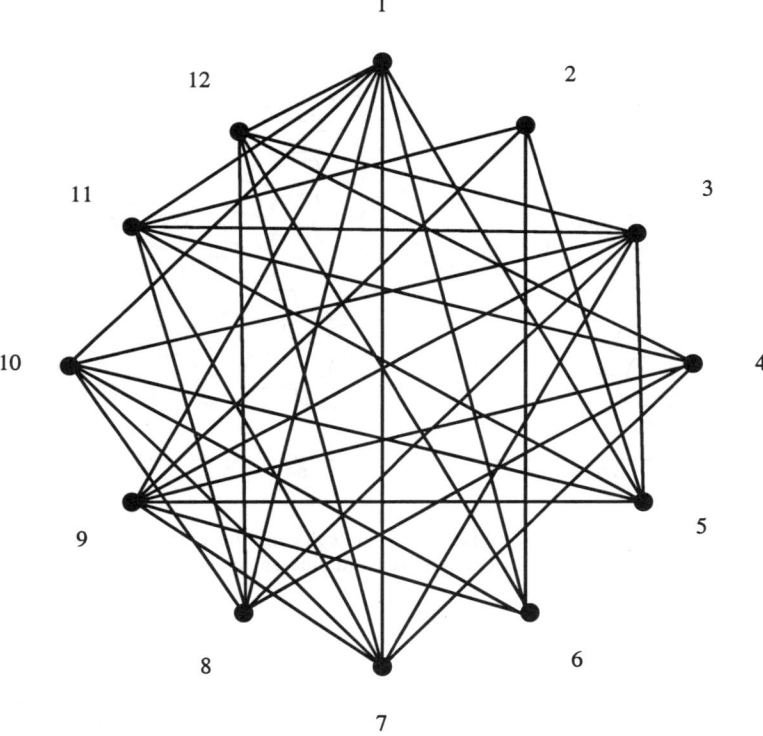

Fig. 5.11   Graph with 12 vertices and chromatic number 3.

ever, the repulsion here is differential and occur in continuous time. In both the antivoter model and the coupled oscillators model, there are theoretical results available for 2-colorable graphs, but none are available for more general graphs. One difference between the two approaches is that the antivoter model apparently always generates a 3-coloring if one waits long enough while the present model doesn't seem to do that. One the other hand, the antivoter model will not return a valid coloring if the algorithm is stopped prematurely or the graph is not 3-colorable, while in the present model, after a fixed initial period it seems the algorithm almost always generates a valid coloring when stopped at any time. Furthermore, the antivoter model for 3-coloring will not return a 2-coloring for 2-colorable graphs, while computer simulations indicate that the present system almost always returns a 2-coloring for 2-colorable graphs.

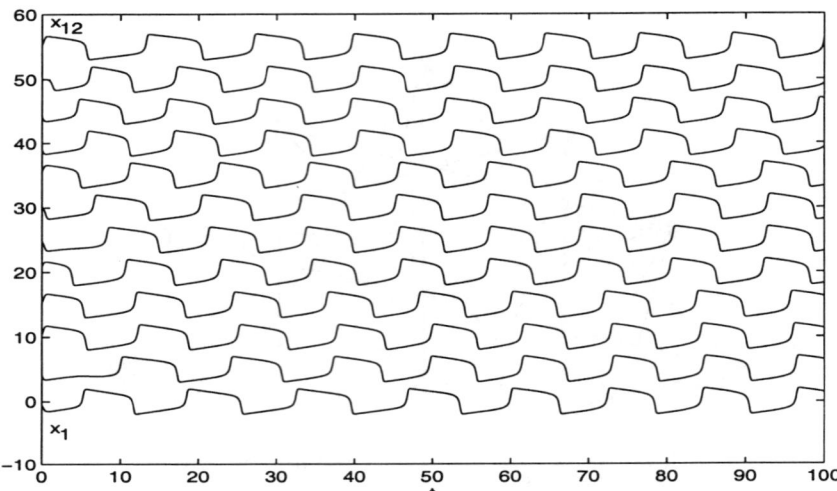

Fig. 5.12 Waveforms corresponding to x_i for coloring the graph in Fig. 5.11. The curves are translated in the vertical axis to avoid clutter. x_1 is the lowest curve and x_{12} is the highest curve.

The system as described does not operate perfectly for graphs which are not 2-colorable. Once in a while, the system would not generate a valid coloring or would converge to a stable equilibrium point, in which case the system has to be simulated using a different set of initial conditions and different G. This behavior requires further investigation. Another future research direction is studying whether the diagonal matrix $\text{diag}(1/v(i))$ in the definition of G can be replaced by a more optimal (not necessarily diagonal) matrix.

5.4.4 Calculating the star chromatic number of a graph

In [96], Vince defined the concept of a *star chromatic number* which is a generalization of the standard chromatic number. Defining Z_k as R/kZ, the k-chromatic number χ_k corresponds to a Z_k-coloring of the graph where the colors are elements of the ring Z_k with the colors of adjacent vertices as far apart as possible. The metric used in Z_k is the circular norm: $|x|_k$

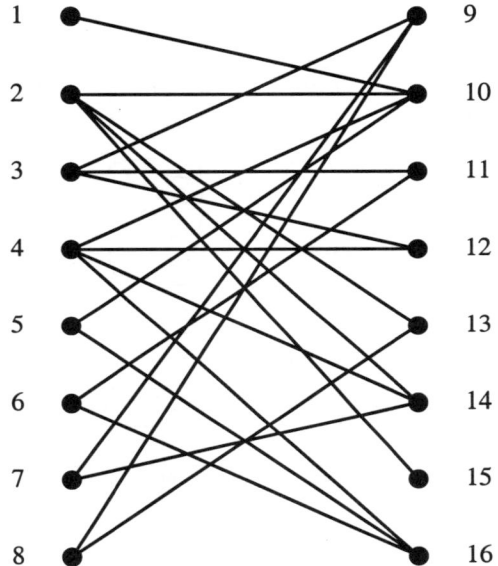

Fig. 5.13 Connected bipartite graph with 16 vertices.

is the distance of x to the nearest multiple of k. χ_k is defined as

$$\chi_k = \min_c \frac{k}{\min_{(i,j)\in E} |c(i) - c(j)|_k}$$

The star chromatic number χ^* is defined as the infimum of the k-chromatic numbers. It was shown in [96] that $\chi - 1 < \chi^* \leq \chi$ where χ is the chromatic number and if the graph is connected with n vertices, then $\chi^* = \min_{1 \leq k \leq n} \chi_k$.

We can think of the phase of the oscillator as elements of Z_k with k normalized to 2π. The function h defined earlier then corresponds to the circular norm $|\cdot|_k$ and the procedure of defining μ_d corresponds to a mapping from a Z_k-coloring to a standard graph coloring. Therefore we conjecture that at a synchronized state of the coupled array which generates the minimal (standard) coloring of the graph, this state also corresponds to a Z_k-coloring of the graph which generates χ^*. Therefore the conjecture implies that we can calculate χ^* as:

$$\frac{2\pi}{\min_{(i,j)\in E} h(\phi(i), \phi(j))} \tag{5.5}$$

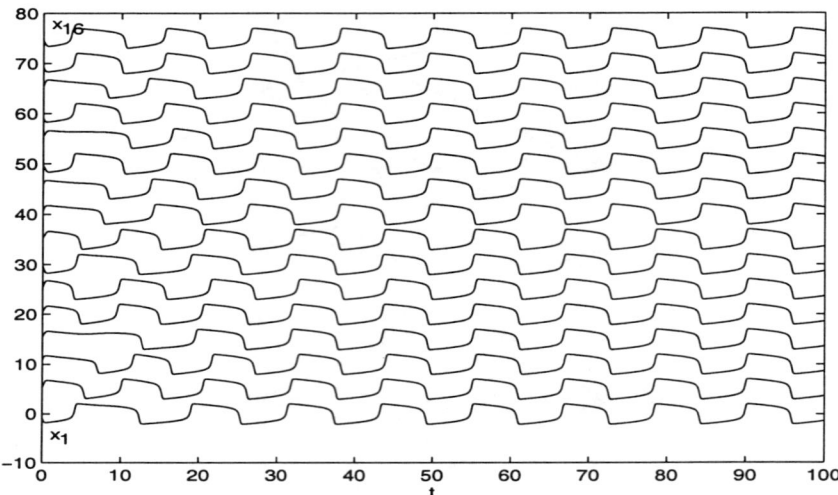

Fig. 5.14 Waveforms corresponding to x_i for coloring the graph in Fig. 5.13. The curves are translated in the vertical axis to avoid clutter. x_1 is the lowest curve and x_{16} is the highest curve.

For bipartite graphs (i.e. graphs with chromatic number 2), the oscillators corresponding to the two colors are at opposite phase at synchronization, and thus Eq. (5.5) evaluates to 2. Thus the conjecture is true for bipartite graphs since $\chi^* = 2$ for bipartite graphs. Let $G_{n,m}$ be defined as the graph with n vertices such that vertex i is adjacent to vertex j if and only if $|i - j|_n \geq m$. For $1 \leq m \leq \frac{n}{2}$, $\chi^*(G_{m,n}) = \frac{n}{m}$ [96]. Initial computer simulations with graphs of the form $G_{n,m}$ support the conjecture.

Chapter 6

Lyapunov Exponents Approach to Synchronization

While the Lyapunov function approach used so far can give mathematically rigorous results, it suffers from one drawback: there are no systematic ways for finding Lyapunov functions for all classes of circuits and systems. On the other hand, the Lyapunov exponents approach [97, 3, 4], even though less rigorous and requires numerical computations and approximations, can be applied to a much larger class of systems. The main difference between the two approaches is that instead of using a Lyapunov function (Lyapunov's second method or direct method) to prove asymptotical stability, the Lyapunov exponents approach calculates Lyapunov exponents (when they exist) of trajectories to test for asymptotical stability. The basic idea is as follows: if the Lyapunov exponents of some trajectory of $\dot{x} = f(x,t)$ are all negative, then $\dot{x} = f(x,t)$ is asymptotically stable around this trajectory. Note however that, unlike the Lyapunov function approach, the Lyapunov exponents approach generally cannot provide global nor robust stability results with respect to parameter variations.

Most of the results discussed so far have a counterpart using the Lyapunov exponents approach. For instance, an analogous result to Theorem 2.1 is:

System (2.1) synchronizes in the sense that $\|x - y\| \to 0$ as $t \to \infty$, if the Lyapunov exponents of $\dot{x} = f(x, u(t), v(t), t)$ are all negative, for all $u(t)$ and $v(t)$ trajectories of $\dot{x} = f(x, x, x, t)$.

In practice, $u(t)$ and $v(t)$ are set to the same single trajectory on or near the chaotic attractor of $\dot{x} = f(x, x, x, t)$.

6.1 Continuous-time systems

Consider the state equations of an array of coupled circuits given by:

$$\dot{x} = (I + p_1(G) \otimes D_1)F((I + p_2(G) \otimes D_2)x) + (p_3(G) \otimes D_3)x \qquad (6.1)$$

where $F(x) = (f(x_1), \ldots, f(x_n))^T = I \otimes f(x_i)$ and p_i are polynomials such that $p_i(0) = 0$ and G is a zero row sum *diagonalizable* matrix.

In the Lyapunov exponents approach, synchronization is deduced from the Lyapunov exponents transverse to the synchronization subspace (we will simply call these the transverse Lyapunov exponents) which are calculated for a trajectory on the synchronization subspace. In particular, we conclude that the system is synchronizing if the transverse Lyapunov exponents are negative for a trajectory on the synchronization subspace.

For such a trajectory $x(t)$, the Jacobian matrix is given by:

$$J(t) = (I + p_1(G) \otimes D_1)(I \otimes Df(t))(I + p_2(G) \otimes D_2) + p_3(G) \otimes D_3 \qquad (6.2)$$

If G is diagonalizable to Λ, then the variational equation $\dot{x} = J(t)x$ is linearly conjugate to

$$\dot{x} = [(I + p_1(\Lambda) \otimes D_1)(I \otimes Df(t))(I + p_2(\Lambda) \otimes D_2) + p_3(\Lambda) \otimes D_3]\, x$$

The matrix

$$(I + p_1(\Lambda) \otimes D_1)(I \otimes Df(t))(I + p_2(\Lambda) \otimes D_2) + p_3(\Lambda) \otimes D_3$$

is block-diagonal with the blocks equal to

$$H(\mu, t) = (I + p_1(\mu)D_1)Df(t)(I + p_2(\mu)D_2) + p_3(\mu)D_3$$

where μ ranges over the eigenvalues of G. The eigenvector $(1, \ldots, 1)^T$ of G corresponds to the synchronization subspace and thus the transverse Lyapunov exponents can be found from the Lyapunov exponents of $\dot{x} = H(\lambda, t)x$ for each eigenvalue λ of G in $L(G)$ (Definition 3.8). Analogous to the set S in Section 3.1, we can define $\tilde{S}$ as the set of λ's such that all Lyapunov exponents of $\dot{x} = H(\lambda, t)x$ are negative. Thus the coupled system is synchronizing if $L(G) \subset \tilde{S}$. Note that since the Jacobian matrix is given by Eq. (6.2) only for trajectories on the synchronization subspace, this synchronization result is only valid locally near the synchronization subspace.

If G has a zero eigenvalue of multiplicity larger than 1, the Lyapunov exponents of $\dot{x} = f(x)$ form a subset of the transverse Lyapunov exponents. This follows from the facts that in this case $0 \in L(G)$ and $H(0,t) = Df(t)$. Thus if in addition f has positive Lyapunov exponents then the array will not synchronize.

Special cases of interest are when two of the polynomials p_i are zero. For example, [98] considers the case when $p_1 = p_2 = 0$ and p_3 is the identity. In this case $H(\lambda, t)$ is simplified to $Df(t) + \lambda D_3$. Generally the relationship between the Lyapunov exponents of $\dot{x} = Df(t)x$ and the Lyapunov exponents of $\dot{x} = (Df(t) + \lambda D_3)x$ is nontrivial. However, when D_3 is a multiple of the identity matrix, it is left as an exercise to show that the Lyapunov exponents of $\dot{x} = Df(t)x$ and the Lyapunov exponents of $\dot{x} = (Df(t) + \lambda I)x$ differ by the constant λ.

6.2 Discrete-time systems

Similarly, the Lyapunov exponents approach can been used to infer synchronization in discrete-time systems.

Consider the following coupled discrete time system:

$$x(k+1) = (I + p_1(G) \otimes D_1)F_k((I + p_2(G) \otimes D_2)x(k)) + (p_3(G) \otimes D_3)x(k) \tag{6.3}$$

where $F_k(x) = (f_k(x_1), \ldots, f_k(x_n))^T = I \otimes f_k(x_i)$, p_i are polynomials such that $p_i(0) = 0$ and G is a *diagonalizable* zero row sum matrix.

For a trajectory $x(k) = (x_1(k), x_1(k), \ldots x_1(k))^T$ of the coupled system on the synchronization manifold, the Lyapunov exponents are given by

$$\lim_{n \to \infty} \ln|\lambda_i(n)|^{\frac{1}{n}} \tag{6.4}$$

where $\lambda_i(n)$ are the eigenvalues of $\Pi_{k=0}^{k=n-1} J(k)$ and $J(k) = (I + p_1(G) \otimes D_1)(I \otimes Df_k(x_1(k)))(I + p_2(G) \otimes D_2) + p_3(G) \otimes D_3$. If C is a matrix which diagonalizes G, then $(C \otimes I)J(k)(C^{-1} \otimes I)$ is block-diagonal and hence $(C \otimes I)\Pi_{k=0}^{k=n-1} J(k)(C^{-1} \otimes I)$ is block-diagonal and the eigenvalues of $\Pi_{k=0}^{k=n-1} J(k)$ are given by the eigenvalues of $H(\mu, n)$ where μ varies over the eigenvalues of G and $H(\mu, n)$ is defined as:

$$H(\mu, n) = \Pi_{k=0}^{k=n-1}\left((I + p_1(\mu)D_1)Df(x_1(k))(I + p_2(\mu)D_2) + p_3(\mu)D_3\right) \tag{6.5}$$

The synchronization manifold A (Definition 5.1) is invariant under Eq. (6.3) and its elements are eigenvectors of $\Pi_{k=0}^{k=n-1} J(k)$. Therefore the Lyapunov exponents of the trajectory $x(k)$ transverse to A are given by Eq. (6.4) where λ_i are the eigenvalues of $H(\mu, n)$ for each $\mu \in L(G)$ (Definition 3.8).

Similar to the continuous-time case, if G has a zero eigenvalue of multiplicity larger than 1, the Lyapunov exponents of f_k will be a subset of the transverse Lyapunov exponents, and thus if f_k has a positive Lyapunov exponent, then the coupled array will not synchronize. This follows from the facts that in this case $0 \in L(G)$ and $H(0, n) = \Pi_{k=0}^{k=n-1} D f_k(x_1(k))$.

The expression for $H(\mu, n)$ can be simplified significantly if $p_3 = 0$ and the discrete-time systems are scalar systems, i.e., D_i are scalars and f_k are real-valued functions of a single variable. In this case $H(\mu, n) = (1 + p_1(\mu) D_1 + p_2(\mu) D_2 + p_1(\mu) p_2(\mu) D_1 D_2)^n \Pi_{k=0}^{k=n-1} f'_k(x_1(k))$. Let $q(\mu) = \ln|1 + p_1(\mu) D_1 + p_2(\mu) D_2 + p_1(\mu) p_2(\mu) D_1 D_2|$. In particular, the transverse Lyapunov exponents are $q(\mu) + \lambda_i$ where $\mu \in L(G)$ and λ_i are the Lyapunov exponents of f_k. Therefore the coupled array is synchronizing if $-q(\mu)$ is bigger than the largest Lyapunov exponent of f_k for all $\mu \in L(G)$.

In [99] this approach was used to study synchronization in the case of arrays of coupled scalar maps where $p_1 = p_3 = 0$, p_2 is the identity and D_2 is a scalar, in which case $q(\mu) = \ln|1 + \mu D_2|$.

Consider Example 3.2 in Section 3.4 where $L(G) = \{\epsilon\}$ and $D_2 = -1$. In this case $-q(\mu) = -\ln|1 - \epsilon|$, i.e. the array synchronizes if $|1 - \epsilon| < e^{-\lambda}$ where λ is the Lyapunov exponent of the logistic map.

6.3 Three oscillator universal probe for determining synchronization in coupled arrays

In [73], Fink et al. describe how a coupled array of three oscillators can serve as a universal probe of the synchronization properties in general coupled arrays of oscillators. The universal probe allows the synchronization in arbitrarily coupled arrays of oscillators to be determined by observing whether the universal probe synchronizes for a discrete set of parameters.

Consider the coupled array of oscillators:

$$\dot{x} = I \otimes f(x_i) + G \otimes H(x_i) \quad (6.6)$$

where $x = (x_1, x_2, \ldots, x_m)^T$ and G is an irreducible matrix with zero row sums. The case where G has a constant row sum γ can be reduced to the case of G having zero row sums by replacing $f(x_i)$ with $f(x_i) + \gamma H(x_i)$.

If G is diagonalizable, the analysis in Section 6.1 shows (noting that D_3 can be a nonlinear function) that the Lyapunov exponents transverse to the synchronization manifold of the coupled array (6.6) can be deduced from the Lyapunov exponents of the variational equations:

$$\dot{\eta} = (Df(z) + \mu DH(z))\eta \qquad (6.7)$$

for each μ a nonzero eigenvalue of G. This leads to the study of a three oscillator probe, which contains variational equations of the form (6.7) for some μ. By varying two real parameters, μ can be set to any complex number. Thus this three oscillator probe can determine the transverse Lyapunov exponents, and hence can determine whether an arbitrarily coupled array synchronizes, by setting the parameter μ to each of the nonzero eigenvalues of G. In particular, if the resulting probe synchronizes for each μ a nonzero eigenvalue of the coupling matrix G, then Eq. (6.6) synchronizes. This is important and useful in practical settings when the exact description of the vector field f is unknown or the determination of Lyapunov exponents via numerical computations is difficult.

The three oscillator probe [73] has the state equations (6.6) with coupling matrix

$$G = \begin{pmatrix} -2\frac{\epsilon}{3} & \frac{\epsilon}{3} + \frac{\delta}{\sqrt{3}} & \frac{\epsilon}{3} - \frac{\delta}{\sqrt{3}} \\ \frac{\epsilon}{3} - \frac{\delta}{\sqrt{3}} & -2\frac{\epsilon}{3} & \frac{\epsilon}{3} + \frac{\delta}{\sqrt{3}} \\ \frac{\epsilon}{3} + \frac{\delta}{\sqrt{3}} & \frac{\epsilon}{3} - \frac{\delta}{\sqrt{3}} & -2\frac{\epsilon}{3} \end{pmatrix} \qquad (6.8)$$

G describes how the three oscillators are coupled to each other. The variational equation contains Eq. (6.7) with $\mu = \epsilon + i\delta$ where $i = \sqrt{-1}$. Fig. 6.1 shows schematically how the oscillators are coupled.

In practical implementations, the probe (6.6) with G defined as (6.8) can cause some difficulties. First of all, the form of G implies that all three oscillators are connected to each other. There are 6 nonzero off-diagonal terms in G, implying 6 coupling terms between the three oscillators. In general, the higher the number of coupling terms, the more complicated the coupled array. Furthermore, the variable parameters ϵ and δ appear in several entries of G, meaning that they need to be varied over some range of parameters and at the same time be perfectly matched to each other in the corresponding coupling terms. In particular, in [73] the exact same value

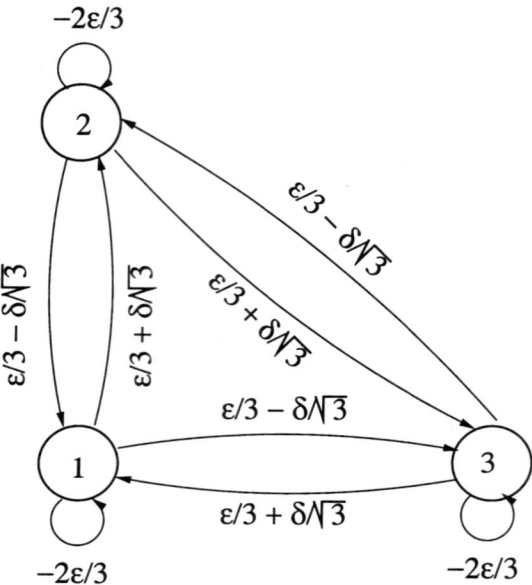

Fig. 6.1 Schematic of coupling scheme for a universal probe reported in [73]. The circles indicate the three oscillators and the labels on the edges are the weights of the corresponding coupling terms.

for ϵ (and for δ) needs to be sent to the inputs of several matching analog multipliers. This matching problem becomes more problematic when the analog multipliers are replaced by physical resistors as is common in several circuit implementations of coupled arrays.

There exists other universal probes for synchronization stability which use the minimum number of coupling terms and minimize the occurrence of the same variable parameter in several coupling terms [100]. The main idea is that there exists other matrices G besides Eq. (6.8) which also allow Eq. (6.6) to be a universal probe. The only requirement is that G is a real-valued diagonalizable 3×3 zero row sum matrix with two real parameters ϵ and δ such that the eigenvalues of G are $\{0, \epsilon + i\delta, \epsilon - i\delta\}$. We will call this *requirement A*. It's clear that we only need to consider $\delta \geq 0$ as the case $\delta < 0$ is taken care of by the conjugate eigenvalues. This explains why all the boundaries between the stable and unstable region in the Master Stability Function plots in [73] are symmetric with respect to the real axis. In fact, more is true. All the specific matrices G considered in this section (Eqs. (6.8)-(6.9), Eqs. (6.11)-(6.14)) have the following property: G corresponding to $\epsilon + i\delta$ is similar to G corresponding to $\epsilon - i\delta$ via a permutation matrix*. In other words, the three oscillator universal probe for $\epsilon + i\delta$ is equivalent to the probe for $\epsilon - i\delta$ after relabeling the oscillators.

Consider the following matrix G:

$$G = \begin{pmatrix} 0 & 0 & 0 \\ \delta - \epsilon & \epsilon & -\delta \\ -\delta - \epsilon & \delta & \epsilon \end{pmatrix} \tag{6.9}$$

Note that this matrix G satisfies requirement A and has 4 nonzero off-diagonal terms. Furthermore, there is no coupling into the first oscillator. Its trajectories are those of an uncoupled oscillator. One possible implementation of this universal probe is to have the first oscillator replaced by a "memory" device which plays back the trajectory of an uncoupled oscillator. A schematic of this coupling scheme is shown in Fig. 6.2. Some elements of G still need to be matched to each other.

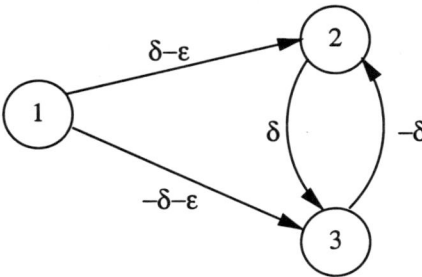

Fig. 6.2 Schematic of coupling scheme for a simplified universal probe.

As mentioned above, we want to minimize the number of coupling elements, which corresponds to the number of nonzero off-diagonal elements of G. What is the minimum number of nonzero off-diagonal elements for matrices G which satisfy requirement A? If a zero row sum matrix G has two nonzero off-diagonal elements, they must be on different rows, as otherwise G will have two zero eigenvalues. In this case G takes on one of the

---

*In fact, for Eqs. (6.12)-(6.14) the matrix G corresponding to $\epsilon + i\delta$ is equal to G corresponding to $\epsilon - i\delta$.

following three forms:

$$\begin{pmatrix} 0 & 0 & 0 \\ 0 & a & -a \\ 0 & b & -b \end{pmatrix}, \begin{pmatrix} 0 & 0 & 0 \\ 0 & a & -a \\ -b & 0 & b \end{pmatrix}, \begin{pmatrix} 0 & 0 & 0 \\ -a & a & 0 \\ -b & 0 & b \end{pmatrix} \quad (6.10)$$

The first matrix has two zero eigenvalues, while the second and third matrices have as eigenvalues $\{0, a, b\}$ which are all real. Therefore G must have at least 3 nonzero off-diagonal elements to satisfy requirement A.

If $\delta = 0$, i.e. the eigenvalue $\mu = \epsilon + i\delta$ is real, and the first oscillator is uncoupled, i.e. the first row of G is zero, then a matrix G satisfying requirement A must be of the form:

$$\begin{pmatrix} 0 & 0 & 0 \\ -\epsilon & \epsilon & 0 \\ -\epsilon & 0 & \epsilon \end{pmatrix}$$

This can be seen as follows. Since G is diagonalizable it can be written as

$$G = V \begin{pmatrix} 0 & 0 & 0 \\ 0 & \epsilon & 0 \\ 0 & 0 & \epsilon \end{pmatrix} V^{-1} = \epsilon V \begin{pmatrix} 0 & 0 & 0 \\ 0 & 1 & 0 \\ 0 & 0 & 1 \end{pmatrix} V^{-1}$$

$$= \epsilon \left(I - V \begin{pmatrix} 1 & 0 & 0 \\ 0 & 0 & 0 \\ 0 & 0 & 0 \end{pmatrix} V^{-1} \right) = \epsilon (I - ab^T)$$

where a is the first column of V and b^T is the first row of V^{-1}. Let $e_1 = (1, 0, 0)^T$ and $e = (1, 1, 1)^T$. Since the first row of G is zero, $e_1^T G = 0$ and thus $e_1^T = e_1^T ab^T$ and thus $b = \frac{1}{e_1^T a} e_1$. Similarly G having zero row sums implies that $Ge = 0$ and thus $a = \frac{1}{b^T e} e$. This implies that $e_1^T a = \frac{1}{b^T e} e_1^T e = \frac{1}{b^T e}$. Therefore $G = \epsilon(I - ab^T) = \epsilon(I - ee_1^T)$ which is what we set out to show.

Note that if $\delta = 0$, a two oscillator probe with coupling matrix

$$G = \begin{pmatrix} 0 & 0 \\ -\epsilon & \epsilon \end{pmatrix} \quad (6.11)$$

suffices as a universal probe.

Let us assume that $\delta \neq 0$. In this case, consider the three oscillator probe (6.6) with coupling matrix:

$$G = \begin{pmatrix} 0 & 0 & 0 \\ 0 & -1 & 1 \\ \epsilon^2 + \delta^2 & -(\epsilon+1)^2 - \delta^2 & 2\epsilon+1 \end{pmatrix} \quad (6.12)$$

Since $\delta \neq 0$, all eigenvalues of G are distinct, G is diagonalizable and satisfies requirement A. G also contains the minimal number of nonzero off-diagonal elements. By using the map $(\alpha, \beta) = (\epsilon^2 + \delta^2, -(\epsilon+1)^2 - \delta^2)$, G can be written as

$$G = \begin{pmatrix} 0 & 0 & 0 \\ 0 & -1 & 1 \\ \alpha & \beta & -\alpha - \beta \end{pmatrix} \quad (6.13)$$

which minimizes the problem of having to match parameter values. Again, the first oscillator operates in the uncoupled mode. A schematic of this coupling scheme is shown in Fig. 6.3.

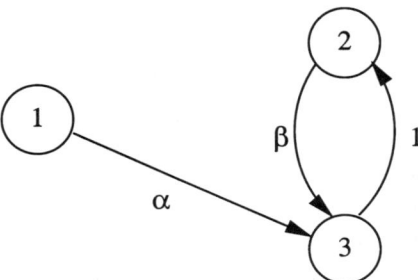

Fig. 6.3 Schematic of coupling scheme for another simplified universal probe for the case when the eigenvalues are not real. In this scheme there are only three coupling elements between oscillators and a minimal number of coupling elements which need to be matched.

Written explicitly, the state equations of this universal probe are:

$$\begin{aligned} \dot{x}_1 &= f(x_1) \\ \dot{x}_2 &= f(x_2) + H(x_3) - H(x_2) \\ \dot{x}_3 &= f(x_3) + \alpha(H(x_1) - H(x_3)) + \beta(H(x_2) - H(x_3)) \end{aligned}$$

To illustrate the simplicity of the coupling matrix (6.13), consider a coupled array of three Chua's oscillators [101] where the coupling occurs between the state variable V_{C2} of each oscillator. The circuit diagram of the array is shown in Fig. 6.4. Note that since $\alpha > 0$, $\beta < 0$, the two coupling resistors with conductances αS and $(\beta - 1)S$ are passive and active respectively. The matching of the term $-\alpha - \beta$ to the terms α and β in Eq. (6.13) is satisfied automatically via Kirchhoff's laws.

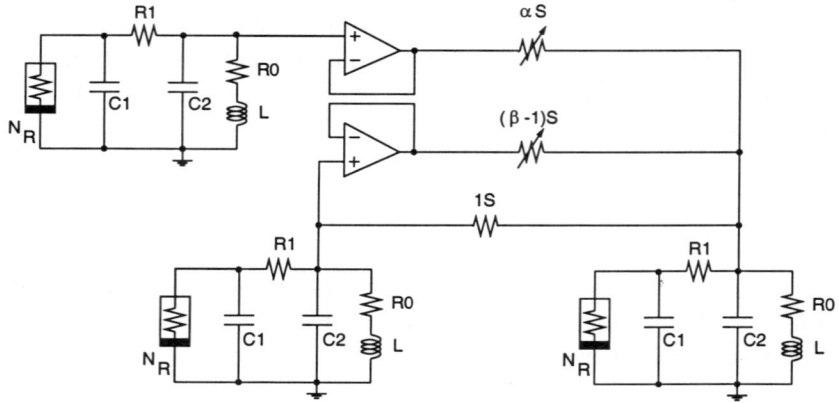

Fig. 6.4 Universal probe consisting of 3 coupled Chua's oscillators. The coupling matrix is given by Eq. (6.13) and the conductances of the coupling resistors are shown.

Similarly, a universal probe for Chua's oscillators using coupling matrix (6.9) can be implemented using a variable gyrator [6] as shown in Fig. 6.5. In some applications, the coupling between Chua's oscillators in a coupled array occurs via passive resistors, resulting in a coupling matrix whose nonzero eigenvalues only have negative real parts, i.e. $\epsilon < 0$. In this case the variable resistors only need to be passive in the universal probe in Fig. 6.5.

Note that the matrix G in Eq. (6.12) has 5 nonzero entries. If $\epsilon \neq 0$ and $\delta \neq 0$, then

$$G = \begin{pmatrix} 0 & 0 & 0 \\ 0 & 2\epsilon & -2\epsilon \\ -\frac{\epsilon^2+\delta^2}{2\epsilon} & \frac{\epsilon^2+\delta^2}{2\epsilon} & 0 \end{pmatrix} \quad (6.14)$$

satisfies requirement A and has only 4 nonzero entries. This is schematically shown in Fig. 6.6. The minimum number of nonzero entries for a matrix G satisfying requirement A is 4 since G must have at least one nonzero diagonal element (as the trace is equal to 2ϵ and is thus generally nonzero)

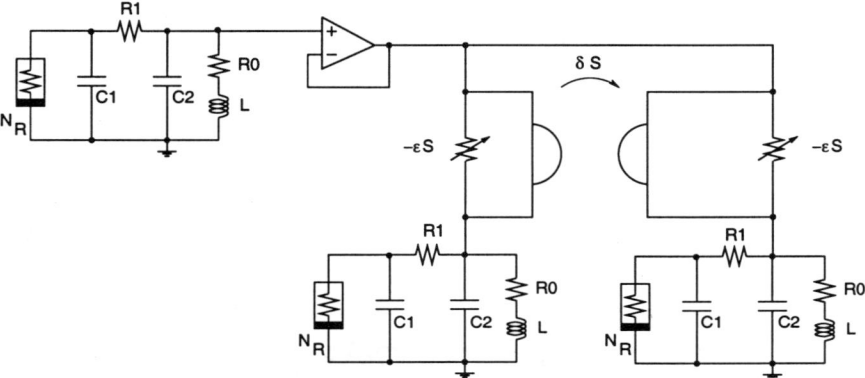

Fig. 6.5 Universal probe consisting of 3 coupled Chua's oscillators. The coupling matrix is given by Eq. (6.9) and the conductances of the coupling resistors and gyrator are shown.

and 3 nonzero off-diagonal elements.

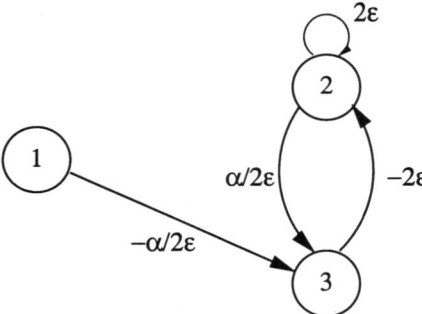

Fig. 6.6 Schematic of a coupling scheme for the case when the eigenvalues are not real nor purely imaginary. In this scheme there are 4 coupling elements in total which is the minimal number of coupling elements for a coupling matrix satisfying requirement A.

Consider a three oscillator probe of the piecewise-linear Rössler systems (Appendix D) with coupling between the x-variables using the coupling matrix in Eq. (6.14). The parameters are $k = 1$, $a = 0.05$, $b = 0.5$, $c = 1$, $f = 0.113$, and $h = 15$. The contour plot of the separation S between the oscillators is shown in Fig. 6.7 which matches quite well with Figs. 6 and 8 in [73].

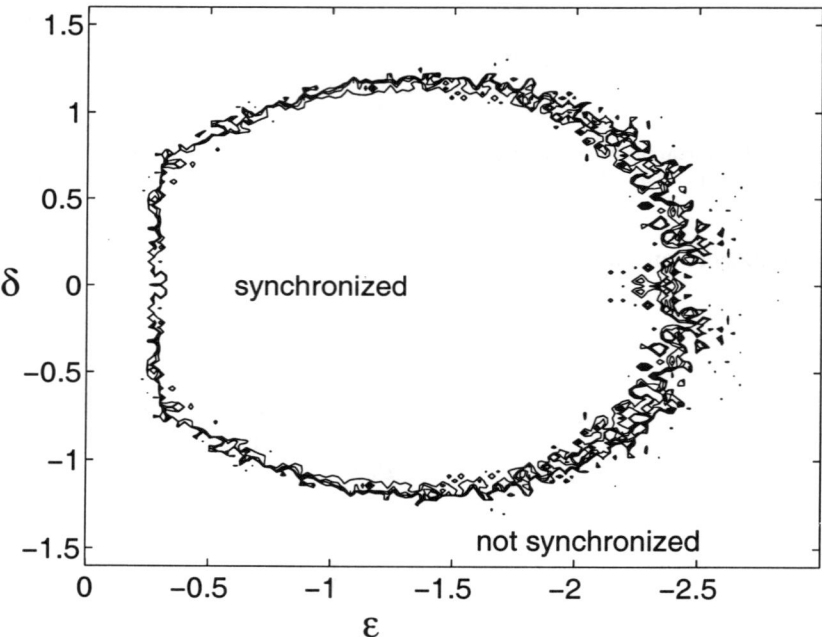

Fig. 6.7 Contour plot of the separation S between oscillators in a universal probe using Eq. (6.14).

Appendix A
Some Linear Systems Theory and Matrix Theory

This appendix lists some of the results on linear algebra, linear systems and matrices we need in this book. Most of the results in this appendix are taken from [13]. We will mainly work with real matrices in this book. Let A^T and A^* denote the transpose and the complex conjugate transpose of the matrix A respectively.

Definition A.1 A real square matrix A is positive definite[†] (semidefinite) if $x^T A x > 0$ ($x^T A x \geq 0$) for all nonzero x. We denote this by $A > 0$ ($A \geq 0$). A square matrix A is negative definite (semidefinite) if $-A$ is positive definite (semidefinite).

A real square matrix A is positive definite (semidefinite) if and only if all eigenvalues of $A + A^T$ are positive (nonnegative). Let I_n denote the n by n identity matrix while I denotes an identity matrix whose size is clear from context.

Definition A.2 A square matrix is Hurwitz if all its eigenvalues have negative real parts.

Definition A.3 A square matrix A is diagonalizable if there exists a nonsingular matrix T such that TAT^{-1} is a diagonal matrix. This implies that A is diagonalizable if and only if A has a full set of eigenvectors.

Definition A.4 The triple of matrices (A, B, C) is a minimal realization of $H(s) = C(sI - A)^{-1}B$ if no smaller size (A, B, C) generates the same $H(s)$.

[†]A is not necessarily symmetric in this definition.

Definition A.5 $H(s)$ is strictly positive real if $\inf_{\omega \in R} \lambda_{\min}(H(j\omega) + H^*(j\omega)) > 0$ where $\lambda_{\min}(A)$ is the smallest eigenvalue of a matrix A whose eigenvalues are all real.

Definition A.6 A matrix A is reducible if there exists a permutation matrix P such that PAP^T is of the form $\begin{pmatrix} B & C \\ 0 & D \end{pmatrix}$ where B and D are square matrices. A matrix is irreducible if it is not reducible.

Lemma A.1 *If the matrices A and B commute and A has constant row sum ϵ and ϵ is a simple eigenvalue of A, then B has constant row sums.*

Proof: Let $e = (1, 1, \ldots, 1)^T$. Then $ABe = BAe = \epsilon Be$. Therefore Be is an eigenvalue of A with eigenvalue ϵ. This implies that $Be = \delta e$ for some δ which implies that B has constant row sum δ. □

Definition A.7 A real square matrix A is called *normal* if $A^T A = AA^T$.

The set of normal matrices corresponds to the set of matrices which are diagonalizable via a unitary matrix [102].

The following lemma lists some properties of real normal matrices which are related to their use in the synchronization in arrays of coupled systems.

Lemma A.2 *Let A be a real normal matrix. Then the following statements are true:*

(A.2.1) If $A + A^T$ is irreducible then A is irreducible.
(A.2.2) If A has constant row sum ϵ then A^T has constant row sum ϵ.
(A.2.3) If $A + A^T$ is a matrix with constant row sum ϵ, such that ϵ is a simple eigenvalue, then A and A^T are matrices with constant row sum $\frac{\epsilon}{2}$.
(A.2.4) If A has only nonnegative off-diagonal elements, then A is irreducible $\Leftrightarrow A + A^T$ is irreducible.

Proof: Suppose that A is normal and reducible. Then A can be written as

$$A = P^T \begin{pmatrix} B & C \\ 0 & D \end{pmatrix} P$$

for some permutation matrix P. Normality of A means that

$$P^T \begin{pmatrix} B^T B & B^T C \\ C^T B & C^T C + D^T D \end{pmatrix} P = P^T \begin{pmatrix} BB^T + CC^T & CD^T \\ DC^T & DD^T \end{pmatrix} P$$

Therefore $CC^T = B^TB - BB^T$. The diagonal elements of B^TB are the inner products of the columns of B. So the trace of B^TB is the sum of the squares of the entries in B. Similarly the trace of BB^T is the sum of the squares of the entries in B. Therefore CC^T has zero trace. Since CC^T is symmetric positive semidefinite, this implies that $CC^T = 0$. Since the diagonal elements of CC^T are the inner products of the rows of C, this means that $C = 0$. So $A + A^T$ is reducible, which proves (A.2.1). Since A is normal, if x is an eigenvector of A with eigenvalue δ, then x is an eigenvector of A^T with eigenvalue $\bar{\delta}$ [66]. Applying this to the eigenvector $(1, 1, \ldots, 1)^T$ yields (A.2.2). Now suppose that $A + A^T$ is a matrix with constant row sum ϵ and a simple eigenvalue ϵ. Since $(A+A^T)$ commute with both A and A^T, by Lemma A.1, A and A^T both has constant row sums. By (A.2.2), it follows that they has the same row sums, which implies that A and A^T have constant row sum $\frac{\epsilon}{2}$, proving (A.2.3).

It's clear that if A has only nonnegative off-diagonal elements, then A being irreducible implies that $A + A^T$ is irreducible since A has no more off-diagonal nonzero entries than $A + A^T$, which proves (A.2.4). □

Next we introduce some results which will be useful in deriving synchronization criteria in coupled arrays.

Lemma A.3 *Let A be an n by n constant row sum matrix. Then the $n - 1$ by $n - 1$ matrix $B = CAG$ satisfies $CA = BC$ where C is the $n - 1$ by n matrix*

$$C = \begin{pmatrix} 1 & -1 & & & \\ & 1 & -1 & & \\ & & \ddots & & \\ & & & 1 & -1 \end{pmatrix} \quad \text{(A.1)}$$

and G is the n by $n - 1$ matrix

$$G = \begin{pmatrix} 1 & 1 & 1 & \cdots & 1 \\ 0 & 1 & 1 & \cdots & 1 \\ & & \ddots & & 1 \\ & & & 1 & 1 \\ 0 & 0 & \cdots & 0 & 1 \\ 0 & 0 & 0 & 0 & 0 \end{pmatrix}$$

Proof: Since A has constant row sums, this implies that CA has zero row sums. The n by n matrix GC is

$$GC = \begin{pmatrix} 1 & 0 & \cdots & 0 & -1 \\ 0 & 1 & 0 & \vdots & -1 \\ 0 & 0 & \ddots & 0 & -1 \\ 0 & 0 & \cdots & 1 & -1 \\ 0 & 0 & \cdots & 0 & 0 \end{pmatrix}$$

Thus the first $n-1$ columns of $CAGC$ is the same as those of CA. The n-th column of $CAGC$ is the negative of the sum of the first $n-1$ columns of CA which is the n-th column of CA since CA has zero row sums. So $CAGC = CA$. The matrix B can be written explicitly as $B_{(i,j)} = \sum_{k=1}^{j} A_{(i,k)} - A_{(i+1,k)}$ for $i, j \in \{1, \cdots, n-1\}$. □

Definition A.8 Let M_1 be the class of matrices C such that row i of C consists of zeros and exactly one entry α_i and one entry $-\alpha_i$ for some nonzero α_i. Let M_2 be the class of matrices $C \in M_1$ such that for any pair of indices i and j there exist indices $i_1, i_2, \ldots, i_l$ with $i_1 = i$ and $i_l = j$ such that for all $1 \leq q < l$, $C_{p,i_q} \neq 0$ and $C_{p,i_{q+1}} \neq 0$ for some p.

Matrices in M_2 can be interpreted in the following way. For $C \in M_1$ construct a graph as follows: the number of vertices of the graph is the number of columns of C, and the number of edges is the number of rows of C. There is an edge between vertex j and vertex l if and only if $C_{i,j} \neq 0$ and $C_{i,l} \neq 0$ for some i. If $C \in M_2$ then this graph is connected. This also implies that the number of columns of $C \in M_2$ is at most one more than the number of rows of C.

There is some relation between these classes of matrices and (node-edge) incidence matrices [6]. For example, a matrix $C \in M_1$ belong to M_2 if after replacing α_i by 1 and $-\alpha_i$ by -1, the resulting matrix is the transpose of the incidence matrix of a *connected* (not necessarily strongly connected) directed graph.

The next several results concern zero row sum matrices and their relation to the matrices in M_1 and M_2. The following lemma gives an upper bound on the second smallest eigenvalue of positive semidefinite zero row sum matrices [84].

Lemma A.4 *Let A be a symmetric positive semidefinite zero row sum matrix. The second smallest eigenvalue λ_2 of A satisfies*

$$\lambda_2 \leq \left(\frac{n}{n-1}\right) \min_i A_{i,i}$$

Proof: Let $e = (1, \ldots, 1)^T$. By the Courant-Fischer theorem [84]

$$\lambda_2 = \min_{x \neq 0, x^T e = 0} \frac{x^T A x}{x^T x}$$

Consider the matrix $B = A - \lambda_2(I - \frac{1}{n}J)$ where J is the matrix of all 1's. Any vector y can be written as $c_1 e + c_2 x$ where $x^T e = 0$. Since B also has zero row sums,

$$y^T B y = c_2^2 x^T B x = c_2^2 (x^T A x - \lambda_2 x^T x) \geq 0$$

The diagonal entries of positive semidefinite matrices are nonnegative [102] and therefore the minimal diagonal entry of B is nonnegative, i.e. $\min_i A_{i,i} - \lambda_2 \left(1 - \frac{1}{n}\right) \geq 0$. □

Definition A.9 *The set W consists of all zero row sum matrices which have only nonpositive off-diagonal elements.*

Definition A.10 *The set W_i consists of all irreducible matrices in W.*

Lemma A.5 *A matrix $A \in W$ satisfies:*

(1) All eigenvalues of A are nonnegative.
(2) 0 is an eigenvalue of A.
(3) If A is irreducible, then 0 is an eigenvalue of multiplicity 1[‡].

Proof: Since A has zero row sums, by Gershgorin's circle criterion all eigenvalues of A are nonnegative. Clearly $(1, \ldots, 1)^T$ is an eigenvector corresponding to the zero eigenvalue. If A is irreducible, then the zero eigenvalue has multiplicity 1 by applying Perron-Frobenius theory [103] to the nonnegative matrix $(\max_i A_{i,i})I - A$. □

For example, if $A + A^T \in W_i$ and A is a real normal matrix, then A and A^T are irreducible zero row sum matrices by Lemmas A.2 and A.5.

Lemma A.6 *If $A \in W_i$ is symmetric and AD is positive semidefinite or negative semidefinite, then D has constant row sums.*

[‡]I.e. 0 is a simple eigenvalue.

Proof: First assume that $AD \geq 0$. Let $e = (1, \ldots, 1)^T$ and $B = AD + D^T A^T$. Then $e^T Be = 0$ since $Ae = 0$. If $Be \neq 0$ then since $B \geq 0$, $e^T Be > 0$. Therefore $Be = 0$ and thus $ADe = 0$. This implies that De is an eigenvector of A with eigenvalue 0, which by Lemma A.5 means that $De = \epsilon e$ for some ϵ. Thus D has constant row sums. The case of $AD \leq 0$ is similar. □

Lemma A.7 *Let A and B be symmetric matrices in W and let $\lambda_2(A)$ denote the second smallest eigenvalue of A. Then $\lambda_2(A) \leq \lambda_2(A+B)$ with the inequality being strict if B is irreducible.*

Proof: Let $e = (1, \ldots, 1)^T$. From the Courant-Fischer theorem, $\lambda_2(A+B)$ can be written as

$$\lambda_2(A+B) = \min_{x \neq 0, x^T e = 0} \frac{x^T(A+B)x}{x^T x}$$

$$\geq \min_{x \neq 0, x^T e = 0} \frac{x^T A x}{x^T x} + \min_{x \neq 0, x^T e = 0} \frac{x^T B x}{x^T x} = \lambda_2(A) + \lambda_2(B)$$

From Lemma A.5, $\lambda_2(B) \geq 0$ with the inequality being strict if B is irreducible. □

Lemma A.8 *If $C \in M_1$, then for any positive integer p, $(M^T M)^p$ is a symmetric zero row sum matrix. A real symmetric matrix A is in W if and only if there exists $C \in M_1$ such that $A = C^T C$. A real symmetric matrix A is in W_i if and only if there exists $C \in M_2$ such that $A = C^T C$.*

Proof: Let $C \in M_1$ and $e = (1, \ldots, 1)^T$. Clearly $(C^T C)^p$ is symmetric for all nonnegative integers p. Since $Ce = 0$, $(C^T C)^p e = 0$ for $p > 0$. So $(C^T C)^p$ has zero row sums if $p > 0$. Now $(C^T C)_{(i,j)}$ is the inner product of columns i and j of C. Let $C \in M_1$. Then the diagonal elements of $C^T C$ is greater than or equal to 0. If $i \neq j$, the only terms in the inner product of columns i and j of C is either 0 or some negative number, so the off-diagonal elements of $C^T C$ is less than or equal to 0. So $C^T C$ has zero row sums and nonpositive off-diagonal elements.

Let A be a symmetric matrix with zero row sums and nonpositive off-diagonal elements. Construct C as follows. For each nonzero row of A we generate several rows of C of the same length as follows: for the i-th row of A, and for each $i < j$ such that $A_{i,j} = -\alpha$ for some $\alpha > 0$, we add a row to C with the i-th element being $\sqrt{\alpha}$, and the j-th element being $-\sqrt{\alpha}$. We claim that this matrix C satisfies $A = C^T C$. First note that since $(1,\ldots,1)C^T = 0$, C^TC is a symmetric zero row sum matrix. Certainly $C \in M_1$. $A_{i,j}$ is the inner product between the i-th column and the j-column of C since by construction, there is only one row of C with nonzero entries in both the i-th and j-th position, giving the appropriate result. From the construction of C, it's clear that A is irreducible if and only if $C \in M_2$. □

The proof of Lemma A.8 gives another characterization of matrices in M_2: a matrix C is in M_2 if and only if $C \in M_1$ and C^TC is irreducible.

If A is a symmetric matrix in W, then A is the node-admittance matrix of a circuit consisting of two terminal resistors with positive conductances [6]. Lemma A.8 is a generalization of the fact that the sum of the Laplacian matrix and the adjacency matrix of a graph is a diagonal matrix with the diagonal entries being the vertex degrees of the vertices.

Lemma A.9 *Let $C \in M_2$ be an $m \times n$ matrix. Then every $n-1$ columns of C are linearly independent. In particular, C has rank $n-1$.*

Proof: C^TC is a symmetric matrix in W_i by Lemma A.8. By Lemma A.5, 0 is a simple eigenvalue of C^TC and therefore C^TC has rank $n-1$. By Sylvester's inequality, C has rank greater than or equal to $n-1$. C has rank $n-1$ since $C(1,1,\cdots,1)^T = 0$, so the null space of C is $\alpha(1,1,\cdots,1)^T$. Therefore any $n-1$ columns of C cannot be linearly dependent. □

A consequence of Lemma A.9 is Kirchhoff's theorem which states that the incidence matrix of a n-node connected directed graph has rank $n-1$ [6].

For a specific matrix C, Lemma A.3 gives an explicit solution to the matrix equation $CA = XC$ for any constant row sum square matrix A. Such a construction exists for a wide class of matrices C which will be of interest to us. In particular, the following lemma shows that we can find a solution to the matrix equation $CA = XC$ for any $C \in M_2$ and A a constant row sum matrix.

Lemma A.10 *Let $C \in M_2$ be an $m \times n$ matrix and A be a constant row sum $n \times n$ matrix. Then there exists a $n \times m$ matrix G_C and B such that $CA = BC$ and $B = CAG_C$, where G_C only depends on C.*

Proof: By Lemma A.9, the first $n-1$ columns of C are linearly independent, i.e. there exist a permutation matrix P such that

$$C = P \begin{pmatrix} C_1 & C_2 \\ C_3 & C_4 \end{pmatrix} = P\tilde{C}$$

where C_1 is a nonsingular $(n-1) \times (n-1)$ submatrix of C. Then $CA = BC$ if and only if $\tilde{C}A = \tilde{B}\tilde{C}$ where $\tilde{B} = P^T BP$. Let $\tilde{A}_1$ denote the first $n-1$ columns of $\tilde{C}A$ and $\tilde{A}_2$ denote the last column of $\tilde{C}A$. The claim is that

$$\tilde{B} = (\ \tilde{A}_1 C_1^{-1} \quad 0\ )$$

satisfies $\tilde{C}A = \tilde{B}\tilde{C}$.

$$\tilde{B}\tilde{C} = (\ \tilde{A}_1 \quad \tilde{A}_1 C_1^{-1} C_2\ )$$

So we need to prove that $\tilde{A}_1 C_1^{-1} C_2 = \tilde{A}_2$. C having zero row sums implies that $(C_1 \quad C_2)$ have zero row sums. This means that

$$C_1^{-1}(C_1 \quad C_2) = (I \quad C_1^{-1} C_2)$$

has zero row sums. Therefore $C_1^{-1} C_2 = (-1, -1, \cdots, -1)^T$. CA has zero row sums and thus $\tilde{C}A$ has zero row sums and therefore

$$\tilde{A}_1 C_1^{-1} C_2 = \tilde{A}_1 (-1, -1, \cdots, -1)^T = \tilde{A}_2$$

The corresponding B is

$$\begin{aligned} B &= P\tilde{B}P^T = P(\tilde{A}_1 C_1^{-1} \quad 0)P^T \\ &= P(\tilde{C}AQC_1^{-1} \quad 0)P^T = (CAQC_1^{-1} \quad 0)P^T \end{aligned}$$

where

$$Q = \begin{pmatrix} & I & \\ 0 & \cdots & 0 \end{pmatrix}$$

Thus $G_C = (QC_1^{-1} \quad 0)P^T$. □

For a fixed C, let us denote the map which maps A into B by S_C: $B = S_C(A) = CAG_C$. Let $\chi_A(\lambda)$ denote the characteristic polynomial of A.

Lemma A.11 *Let C be an $(m-1) \times m$ matrix in M_2 and A be a constant row sum $m \times m$ matrix. If $B = S_C(A)$, then $\chi_A(\lambda) = (\lambda - \epsilon)\chi_B(\lambda)$ and $S_C(A + \alpha I_m) = S_C(A) + \alpha I_{m-1}$. Furthermore, B is the unique $(m-1)$ by $(m-1)$ matrix such that $CA = BC$.*

Proof: First note that following the notation of Lemma A.10

$$CG_C = (CQC_1^{-1} \ 0)P^T = \left(P\begin{pmatrix}C_1\\C_3\end{pmatrix}C_1^{-1} \ 0\right)P^T$$
$$= \left(P\begin{pmatrix}I\\C_3C_1^{-1}\end{pmatrix} \ 0\right)P^T$$

So that when C has order $(m-1) \times m$, C_3 does not exist and $CG_C = I_{m-1}$ and therefore $S_C(A + \alpha I_m) = S_C(A) + \alpha I_{m-1}$.

$$\begin{pmatrix}C & \\ 0 & \cdots & 1\end{pmatrix}(\lambda I - A)\begin{pmatrix}& 1\\ G_C & \vdots\\ & 1\end{pmatrix} = \begin{pmatrix}\lambda I - B & 0\\ & \vdots\\ w & \lambda - \epsilon\end{pmatrix} \quad \text{(A.2)}$$

where w is the last row of $\lambda I - AG_C$. Note that

$$\begin{pmatrix}C & \\ 0 & \cdots & 1\end{pmatrix}\begin{pmatrix}& 1\\ G_C & \vdots\\ & 1\end{pmatrix} = \begin{pmatrix}I & 0\\ * & 1\end{pmatrix}$$

and thus has determinant 1. Taking the determinant of Eq. (A.2) we get $\chi_A(\lambda) = (\lambda - \epsilon)\chi_B(\lambda)$.

If $CA = \tilde{B}C$ then $(B - \tilde{B})C = 0$. Since the rank of C is $m - 1$, by Sylvester's inequality $B - \tilde{B}$ has rank 0 which means $B = \tilde{B}$. This proves uniqueness of B. □

Lemma A.12 *If A is a symmetric matrix in W then A is positive semidefinite and has a zero eigenvalue corresponding to the eigenvector $(1, 1, \ldots, 1)$. If A is irreducible, then the zero eigenvalue has multiplicity 1. Furthermore, A can be decomposed as $A = M^T M$ where M is a matrix in M_1. If A is irreducible, the M can be chosen to be in M_2.*

Proof: A has nonnegative eigenvalues by Lemma A.5. Since A is symmetric, A is positive semidefinite. The rest of this lemma follows from Lemmas A.8 and A.5. □

Appendix B
Graph Theoretical Definitions and Notations

A *hypergraph* $\mathcal{G}$ is a pair $\mathcal{G} = (V, E)$ where V is a set of *vertices*, and E is a set of edges where each edge $e \in E$ is a non-empty subset of V. We can think of the vertices in an edge e to be the vertices e is connected to. The vertex degree of $v \in V$ is the number of edges connected to v. The edge degree of $e \in E$ is the number of elements in e. The (vertex-edge) incidence matrix E of a hypergraph is defined as $E_{ij} = 1$ if $v_i \in e_j$ and 0 otherwise, where v_i and e_j are the i-th vertex and the j-th edge respectively. Thus any $0-1$ matrix without zero columns is the incidence matrix of a hypergraph. If a hypergraph $\mathcal{G}$ has incidence matrix E then the *dual* hypergraph of $\mathcal{G}$ has incidence matrix E^T (when E^T is also a $0-1$ matrix without zero columns). If all vertices have the same degree, then the hypergraph is *regular*. If all edges have the same degree, then the hypergraph is *uniform*. A graph is k-regular (k-uniform) if all vertices (edges) have degree k. A *graph* is a uniform hypergraph where all edges have edge degree 2. The *adjacency matrix* A of a graph is defined as $A_{ij} = 1$ if $i \neq j$ and vertex i is adjacent to vertex j and $A_{ij} = 0$ otherwise. The *Laplacian matrix* and the *algebraic connectivity* of a hypergraph with incidence matrix E are defined as $L = 2(D_v - ED_e^{-1}E^T)$ and the smallest nonzero eigenvalue of L respectively where D_v and D_e are diagonal matrices with the degrees of the vertices and the edges respectively on the diagonal. They were introduced in [104, 70]. If the hypergraph is not connected, the algebraic connectivity is defined as 0. The algebraic connectivity is used to quantify the connectedness of a graph. The constant factor 2 is introduced in the definition of L so that these definitions of Laplacian matrix and algebraic connectivity when restricted to graphs are equivalent to the definitions generally used in the literature [84]§. For a graph, the Laplacian matrix can also be written as $D_v - A$. A (vertex) coloring of a hypergraph $\mathcal{G}$ is an assignment of colors to the vertices of $\mathcal{G}$ such that no two adjacent vertices are given the same color. An edge coloring of $\mathcal{G}$ is a coloring of the dual of $\mathcal{G}$. A coloring of $\mathcal{G}$ with k colors is a k-coloring. A k-colorable graph is a graph which can be colored with k colors. The chromatic number of a graph is the least number of colors it can be colored with. The distance between two vertices is the length of the shortest path between them. The diameter of a graph is the longest distance among pairs of vertices. The mean distance of a graph is the average of all distances between two distinct vertices. The isoperimetric number of a graph $\mathcal{G}$ is defined as $\min_{|X| \leq \frac{n}{2}} \frac{|\delta X|}{|X|}$ where X is a subset of vertices, $|X|$ is the cardinality of X and δX is the set of edges that separates X from the rest of the vertices. A cutset of a connected graph is a minimal set of edges which separates the graph into two disjoint pieces. The edge connectivity of a graph is the size of the smallest cutset of the graph.

We will draw hypergraphs in the following way. The vertices are drawn as closed circles, while edges are denoted as open circles along with branches emanating from these circles connecting the vertices in that edge. For examples, the hypergraph with $V = \{1, 2, 3, 4\}$ and $E = \{\{1, 2\}, \{1, 3, 4\}, \{1, 2, 3\}\}$ is shown in Fig. B.1.

Lemma B.1 *If L is the Laplacian matrix of a (hyper)graph, then L*

- *is symmetric*
- *has zero row sums,*
- *has a zero eigenvalue,*
- *is positive semidefinite,*
- *and has nonpositive off-diagonal elements.*

If the (hyper)graph can be decomposed into k connected components, then L has a zero eigenvalue of multiplicity k.

Proof: Clearly L is symmetric. Suppose that E is an $n \times m$ matrix. Let e_n be the n-vector $(1, \ldots, 1)^T$. Then $E^T e_n = D_e e_m$ and $E e_m = D_v e_n$.

---

§Another way the Laplacian matrix of a hypergraph can be defined is to convert hypergraphs into graphs by replacing hyperedges with weighted cliques of edges connecting the corresponding vertices. By choosing appropriate weights, the Laplacian matrix of the resulting weighted graph is the same as the Laplacian matrix of the hypergraph defined here.

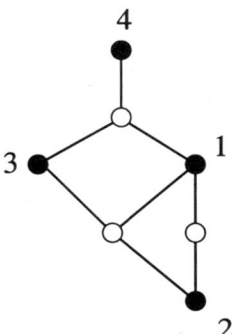

Fig. B.1 Hypergraph with $V = \{1,2,3,4\}$ and $E = \{\{1,2\},\{1,3,4\},\{1,2,3\}\}$.

This implies that $Le_n = 0$ and L has zero row sums. Since E and D_e has nonnegative entries, $ED_e^{-1}E^T$ has nonnegative entries and thus L has nonpositive off-diagonal entries. By Lemma A.12, L has a zero eigenvalue and is positive semidefinite. If the hypergraph can be decomposed into k connected component, then L can be decomposed into k irreducible components, each of which has an zero eigenvalue of multiplicity 1 by Lemma A.12. □

Therefore the Laplacian matrix belongs to the set W (Definition A.9). If the the graph is connected, then the Laplacian matrix belongs to the set W_i (Definition A.10).

Appendix C

Stability, Lyapunov's Direct Method and Lyapunov Exponents

Many of the synchronization results in this book rely on two classes of techniques for determining stability of dynamical systems. The first technique constructs Lyapunov functions to apply Lyapunov's Direct Method. The second technique relies on the computation of Lyapunov exponents.

Let us denote the solution of $\dot{x} = f(x,t)$ with initial conditions x_0 at t_0 as $x(t, x_0, t_0)$. Similarly, the solution of $x(k+1) = f(x(k), k)$ with initial condition x_0 at k_0 is $x(k, x_0, k_0)$.

C.1 Lyapunov function and Lyapunov's direct or second method

In this section we list some basic Lyapunov stability theorems which we need for proving synchronization in coupled circuits and systems. The proofs are adapted from [8, 13] where local versions of these results can be found.

First we need the following definitions from [105, 21].

Definition C.1 A system $\dot{x} = f(x,t)$ (or $x(k+1) = f(x(k), k)$) is asymptotically stable if

(1) For all $\mu > 0$ there exists $\delta > 0$ such that if $\|x(0) - \tilde{x}(0)\| \leq \delta$, then $\|x(t) - \tilde{x}(t)\| \leq \mu$ for all $t \geq 0$ ($\|x(k) - \tilde{x}(k)\| \leq \mu$ for all $k \geq 0$).
(2) For all initial conditions $x(0)$ and $\tilde{x}(0)$, $\|x(t) - \tilde{x}(t)\| \to 0$ as $t \to \infty$, ($\|x(k) - \tilde{x}(k)\| \to 0$ as $k \to \infty$).

Definition C.2 A function $\alpha : R \to R$ is said to belong to class K if

(1) $\alpha(\cdot)$ is continuous and nondecreasing,
(2) $\alpha(0) = 0$,
(3) $\alpha(p) > 0$ whenever $p > 0$.

A basic technique for proving asymptotical stability of $\dot{x} = f(x,t)$ is by Lyapunov's direct method. In this method a nonnegative scalar valued Lyapunov function $V(t,x)$ is constructed and V is shown to decrease along the trajectories of $\dot{x} = f(x,t)$. We assume that all Lyapunov functions we consider are continuous. For a Lyapunov function $V(t,x)$, the generalized derivative along the trajectories of system $\dot{x} = f(x,t)$ is defined as:

$$\dot{V}(t,x) = \lim_{h \to 0^+} \sup \frac{1}{h}[V(t+h, x+hf(x,t)) - V(t,x)]$$

Theorem C.1 *Let μ_1 and μ_2 be continuous real-valued functions. Suppose that a Lyapunov function $V(t,x,y)$, Lipschitz in x,y exists on such that for all $t \geq t_0$, x, y,*

$$a(\|x-y\|) \leq V(t,x,y) \leq b(\|x-y\|)$$

where $a(\cdot)$ and $b(\cdot)$ are functions in class K. Suppose that there exists $\mu > 0$ such that for all $t \geq t_0$ and $\|x-y\| \geq \mu$,

$$\dot{V}(t,x,y) \leq -c$$

for some constant $c > 0$ where $\dot{V}(t,x,y)$ is the generalized derivative of V along the trajectories of

$$\begin{aligned} \dot{x} &= f(x,t) \\ \dot{y} &= \tilde{f}(y,t) \end{aligned} \quad \text{(C.1)}$$

If there exists $\delta > 0$ such that $a(\delta) > b(\mu)$, then for each x_0 and y_0 there exists $t_1 \geq t_0$ such that for all $t \geq t_1$,

$$\|x(t, x_0, t_0) - y(t, y_0, t_0)\| \leq \delta$$

Furthermore, if $\|x_0 - y_0\| \leq \mu$ then

$$\|x(t, x_0, t_0) - y(t, y_0, t_0)\| \leq \delta$$

for all $t \geq t_0$.

Proof: If $V(t, x(t), y(t)) > b(\mu)$ for all $t \geq t_0$ then $b(\|x-y\|) > b(\mu)$ and thus $\|x-y\| > \mu$ for all $t \geq t_0$. This implies that $\dot{V}(t,x,y) \leq -c < 0$ for all $t \geq t_0$

which contradicts the fact that $V(t, x(t), y(t)) \geq 0$. Thus there exists $t_1 \geq t_0$ such that $V(t_1, x(t_1), y(t_1)) \leq b(\mu)$. Now we show that $V(t, x(t), y(t)) \leq b(\mu)$ for all $t \geq t_1$. By way of contradiction, suppose that there exists $t_2 > t_1$ such that $V(t_2, x(t_2), y(t_2)) > b(\mu)$. Then there exists $\epsilon > 0$ such that $V(t_2, x(t_2), y(t_2)) > b(\mu) + \epsilon$. By continuity of $V(t, x(t), y(t))$ with respect to t, there exists $t \in [t_1, t_2)$ such that $V(t, x(t), y(t)) = b(\mu) + \epsilon$. Let

$$t_3 = \sup\{t \in [t_1, t_2) : V(t, x(t), y(t)) = b(\mu) + \epsilon\}$$

Then $V(t_3, x(t_3), y(t_3)) = b(\mu) + \epsilon \leq b(\|x(t_3) - y(t_3)\|)$. Therefore $\|x(t_3) - y(t_3)\| > \mu$ and $\dot{V} \leq -c < 0$ at t_3. So there exists t_4 such that $t_3 < t_4 < t_2$ and

$$V(t_4, x(t_4), y(t_4)) < b(\mu) + \epsilon$$

Therefore there exists $t_5 \in (t_4, t_2)$ such that $V(t_5, x(t_5), y(t_5)) = b(\mu) + \epsilon$ contradicting the fact that t_3 is the largest such t.

Thus we have $a(\|x - y\|) \leq V(t, x(t), y(t)) \leq b(\mu) < a(\delta)$ for all $t \geq t_1$. Therefore $\|x - y\| \leq \delta$ for all $t \geq t_1$. Furthermore, if $\|x_0 - y_0\| \leq \mu$ then $V(t_0, x(t_0), y(t_0)) \leq b(\mu)$ and thus for all $t \geq t_0, V(t, x(t), y(t)) \leq b(\mu)$ which implies that $\|x - y\| \leq \delta$ for all $t \geq t_0$. □

Theorem C.2 *Consider system (C.1) with $\tilde{f} = f$. Suppose that a Lyapunov function $V(t, x, y)$, Lipschitz in x and y, exists such that for all $t \geq t_0$, x, y,*

$$a(\|x - y\|) \leq V(t, x, y) \leq b(\|x - y\|)$$

where $a(\cdot)$ and $b(\cdot)$ are in class K, and for all $t \geq t_0$,

$$\dot{V}(t, x, y) \leq -c(\|x - y\|)$$

for some function $c(\cdot)$ in class K where $\dot{V}(t, x, y)$ is the generalized derivative of V along the trajectories of Eq. (C.1). Then the system $\dot{x} = f(x, t)$ is asymptotically stable.

Proof: Note that for each $\delta > 0$, $0 < a(\delta) \leq b(\delta)$, so there exists $\mu > 0$ such that $a(\delta) > b(\mu)$. Therefore $\dot{V}(t, x, y) \leq -c(\mu) < 0$ for all $\|x - y\| \geq \mu$, and for each $\delta > 0$ there exists by theorem C.1 a time $t_1 > t_0$ such that

$$\|x(t, x_0, t_0) - y(t, y_0, t_0)\| \leq \delta$$

for all $t \geq t_1$. Furthermore, if $\|x_0 - y_0\| \leq \mu$ then

$$\|x(t, x_0, t_0) - y(t, y_0, t_0)\| \leq \delta$$

for all $t \geq t_0$. □

The reader is referred to [106] for generalizations of such Lyapunov theorems.

Definition C.3 System $\dot{x} = f(x, u(t), t)$ is said to be *u-asymptotically stable* if for $x_1(t)$ and $x_2(t)$ trajectories of $\dot{x} = f(x, u_1(t), t)$ and $\dot{x} = f(x, u_2(t), t)$ respectively and $u_1 \to u$, $u_2 \to u$ as $t \to \infty$, we have $x_1 \to x_2$ as $t \to \infty$.

Clearly u-asymptotically stability implies asymptotical stability. The definition of u-asymptotical stability relates to synchronization and stability under noise which eventually dies down. With quadratic Lyapunov functions, we can show that $\dot{x} = f(x,t) + u(t)$ is u-asymptotically stable if $\dot{x} = f(x,t)$ is asymptotically stable [13].

Theorem C.3 *Suppose that a quadratic Lyapunov function $V(t, x, y) = \frac{1}{2}(x - y)^T V (x - y)$ exists where V is a symmetric positive definite matrix, and such that*

$$\dot{V}(t, x, y) \leq -\|x - y\|^2$$

for all $t \geq t_0$, where $\dot{V}(t, x, y)$ is the generalized derivative of V along the trajectory of Eq. (C.1). If $\|u_1(t) - u_2(t)\| \leq \frac{\delta}{cond(V)\|V\|} - \epsilon$ for all $t \geq t_0$ and some $\epsilon > 0$, then there exists $t_1 \geq t_0$ such that the trajectory $(x(t), y(t))$ of system

$$\begin{aligned} \dot{x} &= f(x, t) + u_1(t) \\ \dot{y} &= \tilde{f}(y, t) + u_2(t) \end{aligned} \quad (C.2)$$

satisfies $\|x(t) - y(t)\| \leq \delta$ for all $t \geq t_1$. Furthermore, if $\|x(t_0) - y(t_0)\| \leq \frac{\delta}{cond(V)}$, then $\|x(t) - y(t)\| \leq \delta$ for all $t \geq t_0$. $cond(V) = \|V\|\|V^{-1}\|$ denotes the condition number of V.

Proof: The generalized derivative of V along the trajectories of system Eq. (C.2) satisfies

$$\begin{aligned} \dot{V}(t, x, y) &\leq -\|x - y\|^2 + (x - y)^T V(u_1 - u_2) \\ &\leq -\|x - y\|^2 + \|V\|\|x - y\|\|u_1 - u_2\| \\ &= \|x - y\|(\|V\|\|u_1 - u_2\| - \|x - y\|) \end{aligned} \quad (C.3)$$

and the result follows from Theorem C.1 where we choose $\mu = \frac{\delta}{\text{cond}(V)}$. □

Corollary C.1 *Suppose that a quadratic Lyapunov function $V(t,x,y) = \frac{1}{2}(x-y)^T V(x-y)$ exists where V is a symmetric positive definite matrix, and such that*

$$\dot{V}(t,x,y) \leq -\|x-y\|^2$$

for all $t \geq t_0$, where $\dot{V}(t,x,y)$ is the generalized derivative of V along the trajectory of

$$\begin{aligned} \dot{x} &= f(x,t) \\ \dot{y} &= f(y,t) \end{aligned}$$

Then $\dot{x} = f(x,t) + u(t)$ is u-asymptotically stable.

The above theorems are useful in deriving synchronization results of two coupled systems. Theorems useful for synchronization of an array of coupled systems can be obtained by replacing the $\|x-y\|$ term by a suitable term to quantify the synchronization error between systems in an array.

Theorem C.4 *Consider the coupled array*

$$\begin{aligned} \dot{x}_1 &= f_1(x_1, t) \\ \dot{x}_2 &= f_2(x_2, t) \\ &\vdots \\ \dot{x}_n &= f_n(x_n, t) \end{aligned} \quad (C.4)$$

where $x = (x_1, \ldots x_n)^T$. Suppose that a Lyapunov function $V(t,x)$, Lipschitz in x, exists such that for all $t \geq t_0$, x,

$$a(x^T A x) \leq V(t,x) \leq b(x^T A x)$$

where $a(\cdot)$ and $b(\cdot)$ are in class K, and A is a symmetric matrix in W_i. If for all $t \geq t_0$,

$$\dot{V}(t,x) \leq -c(x^T A x)$$

for some function $c(\cdot)$ in class K where $\dot{V}(t,x)$ is the generalized derivative of V along the trajectories of Eq. (C.4). Then $x_i \to x_j$ as $t \to \infty$ for all i, j.

Proof: By replacing $\|x - y\|$ with $x^T A x$ in Theorems C.1 and C.2, we can show that $\|x^T A x\| \to 0$. By Lemma A.12 $A = C^T C$, $C \in M_2$ and this means that $\|Cx\| \to 0$ which implies that $\|x_i - x_j\| \to 0$ as $t \to \infty$. □

Theorem C.5 *Suppose that a quadratic Lyapunov function $V(t,x) = \frac{1}{2} x^T C^T V C x$ exists where V is a symmetric positive definite matrix and $C \in M_2$, and such that*

$$\dot{V}(t,x) \leq -\|Cx\|^2$$

for all $t \geq t_0$, where $\dot{V}(t,x)$ is the generalized derivative of V along the trajectory of Eq. (C.4). If $\|u_i(t) - u_j(t)\| \to 0$ as $t \to \infty$, Then the trajectories of

$$\begin{aligned}
\dot{x}_1 &= f_1(x_1, t) + u_1(t) \\
\dot{x}_2 &= f_2(x_2, t) + u_2(t) \\
&\vdots \\
\dot{x}_n &= f_n(x_n, t) + u_n(t)
\end{aligned} \quad (C.5)$$

satisfy $\|x_i - x_j\| \to 0$.

C.2 Lyapunov exponents

The Lyapunov exponents of $\dot{x} = f(x,t)$ and $x(k+1) = f(x(k), k)$ measure average local contraction and expansion of the phase space and are defined by:

Definition C.4 *The Lyapunov exponents of $\dot{x} = f(x,t)$ and initial condition x_0 at t_0 are defined by*

$$\lambda_i = \lim_{t \to \infty} \frac{1}{t} \ln \|m_i(t)\|$$

when the limit exists, where $\{m_i(t)\}$ are the eigenvalues of $\frac{\partial x(t, x_0, t_0)}{\partial x_0}$.

Definition C.5 *The Lyapunov exponents of $x(k+1) = f(x(k), k)$ and initial condition $x(k_0) = x_0$ are defined by*

$$\lambda_i = \lim_{k \to \infty} \|m_i(k)\|^{\frac{1}{k}}$$

when the limit exists, where $\{m_i(k)\}$ are the eigenvalues of $\frac{\partial x(k, x_0, k_0)}{\partial x_0}$.

Numerical algorithms for approximating Lyapunov exponents can be found in [107].

Appendix D
Chaotic Circuits and Systems

Most of the examples of coupled arrays of continuous time chaotic systems in this book use the following circuits and systems as the underlying chaotic systems.

D.1 Nonautonomous chaotic circuits and systems

D.1.1 Circuit 1

This circuit is a simple nonautonomous second order system which exhibits chaotic behavior under suitable periodic driving [36] and circuit parameters. The circuit diagram is shown in Fig. D.1.

The state equations for this circuit are given by:

$$\begin{aligned} \frac{dv_1}{dt} &= \frac{1}{C}\left(i_2 - f_c(v_1 + s_c(t))\right) \\ \frac{di_2}{dt} &= -\frac{1}{L}(v_1 + i_2 R) \end{aligned} \quad (D.1)$$

where $s_c(t) = A_c \sin(\Omega t)$ is the periodic forcing function and the v-i characteristic of the voltage-controlled Chua's diode $f_c(v)$ is an odd-symmetric 3-segment piecewise-linear function given by

$$f_c(v) = G_b v + \frac{1}{2}(G_a - G_b)\left(|v + E| - |v - E|\right) \quad (D.2)$$

where $E > 0$ (Fig. D.2).

After normalization using $G = \frac{1}{R}$, $x = \frac{v_1}{E}$, $y = \frac{i_2}{GE}$, $\tau = \frac{t}{|C/G|}$, $a = \frac{G_a}{G}$, $b = \frac{G_b}{G}$, $\omega = \Omega|C/G|$, $\beta = \frac{C}{LG^2}$, $s(t) = \frac{s_c(|C/G|t)}{E}$, $A = \frac{A_c}{E}$, and redefining

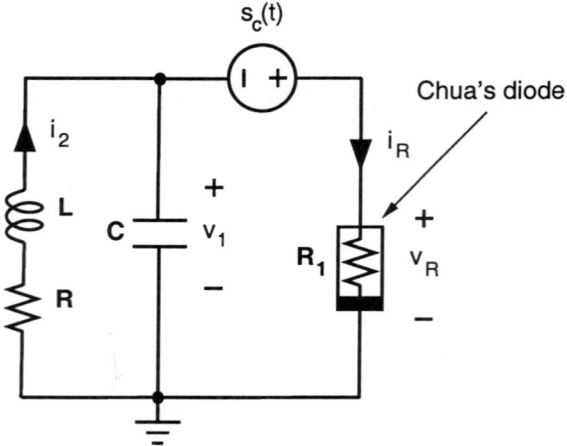

Fig. D.1 Nonautonomous chaotic circuit. There exists parameter for R, C and L passive and R_1 active with a monotone v-i characteristic such that the circuit is chaotic.

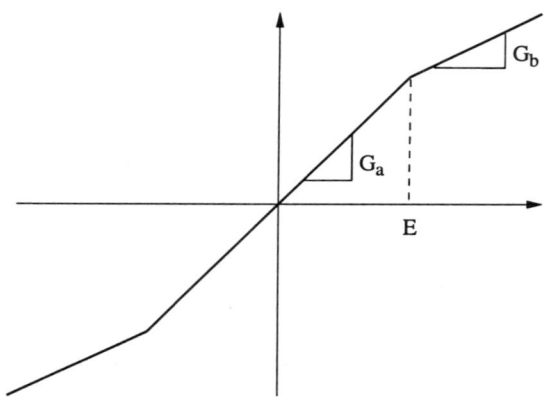

Fig. D.2 3-segment piecewise-linear characteristic of Chua's diode.

τ as t, we obtain the following dimensionless equations:

$$\begin{aligned} \frac{dx}{dt} &= k(y - f(x + s(t))) \\ \frac{dy}{dt} &= k\beta(-x - y) \end{aligned} \quad (D.3)$$

where $k = 1$ if $\frac{C}{G} > 0$ and $k = -1$ if $\frac{C}{G} < 0$, $s(t) = A\sin(\omega t)$ and

$$f(x) = bx + \frac{1}{2}(a - b)\left(|x + 1| - |x - 1|\right) \tag{D.4}$$

For a set of parameters for R, C and L passive and R_1 active with a monotone $v - i$ characteristic, and the driving a periodic sinusoidal signal, the circuit is chaotic. For instance, for the following set of parameters: $a = -1.37$, $b = -0.84$, $\omega = 0.4$, $A = 0.5$, $\beta = 0.895$, and $k = 1$. A chaotic attractor in the x-y plane is shown in Fig. D.3.

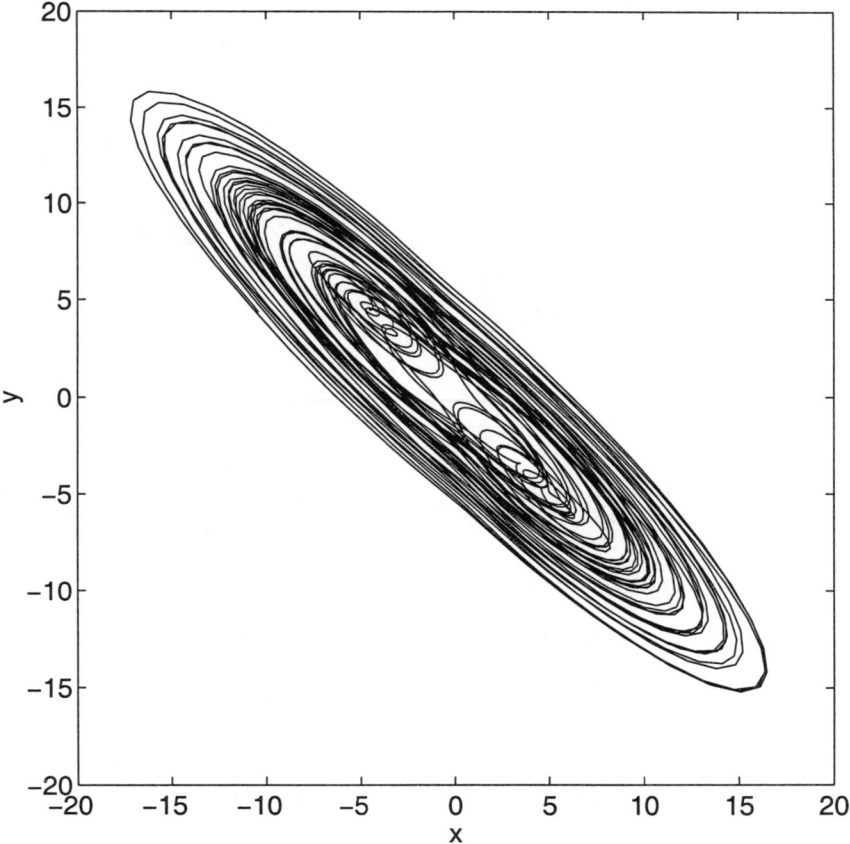

Fig. D.3 Chaotic attractor for system (D.3) in the x-y plane. The parameters are $a = -1.37$, $b = -0.84$, $\omega = 0.4$, $A = 0.5$, $\beta = 0.895$, and $k = 1$.

D.1.2 Circuit 2

The second nonautonomous chaotic circuit is shown in Fig. D.4. It is obtained from Fig. D.1 by interchanging the linear resistor and the nonlinear resistor. It can also be obtained as the dual circuit of the circuit in [108] except that we replace the current source in the dual circuit by a Thévenin equivalent voltage source.

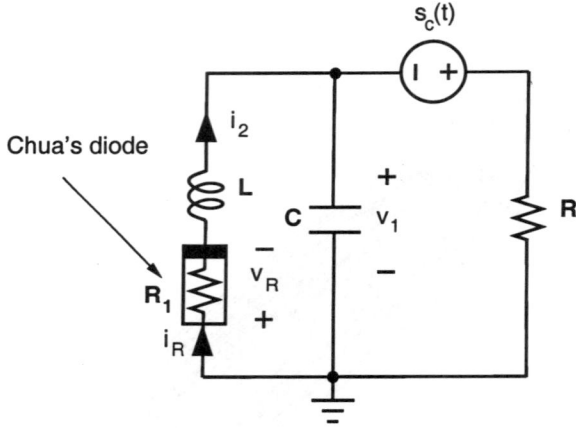

Fig. D.4 Nonautonomous chaotic circuit 2. For the system to become chaotic, we use an active linear resistor and a Chua's diode with a passive monotone v-i characteristic.

The state equations for this circuit are given by:

$$\begin{aligned} \frac{dv_1}{dt} &= \frac{1}{C}\left(i_2 - \frac{1}{R}(v_1 + s_c(t))\right) \\ \frac{di_2}{dt} &= -\frac{1}{L}\left(v_1 + g_c(i_2)\right) \end{aligned} \quad \text{(D.5)}$$

where $s_c(t) = A_c \sin(\Omega t)$ is the periodic forcing function and the v-i characteristic of the current-controlled Chua's diode $g_c(i)$ is again an odd-symmetric 3-segment piecewise-linear function given by

$$g_c(i) = R_b i + \frac{1}{2}(R_a - R_b)\left(|i + I| - |i - I|\right) \quad \text{(D.6)}$$

where $I > 0$.

After normalization using $G = \frac{1}{R}$, $x = \frac{v_1 G}{I}$, $y = \frac{i_2}{I}$, $\tau = \frac{t}{|C/G|}$, $a = GR_a$, $b = GR_b$, $\omega = \Omega |C/G|$, $\beta = \frac{C}{LG^2}$, $s(t) = \frac{s_c(|C/G|t)G}{I}$, $A = \frac{A_c G}{I}$, and redefining τ as t, we obtain the following dimensionless equations:

$$\begin{aligned} \frac{dx}{dt} &= k(y - x - s(t)) \\ \frac{dy}{dt} &= k\beta(-x - f(y)) \end{aligned} \tag{D.7}$$

where $k = 1$ if $\frac{C}{G} > 0$ and $k = -1$ if $\frac{C}{G} < 0$, $s(t) = A\sin(\omega t)$ and f is as defined in Eq. (D.4).

There exists a set of circuit parameter values such that the inductor and the capacitor are passive, the nonlinear resistor has a passive monotone v-i characteristic and the linear resistor is active, i.e. C, L, R_a, $R_b > 0$, and $R < 0$.¶

For example, the chaotic attractor for the following set of parameters: $a = -1.27$, $b = -0.68$, $\omega = 0.5$, $A = 0.2$, $\beta = 1.4$, and $k = -1$. is shown in the x-y plane in Fig. D.5.

D.2 Autonomous chaotic circuits and systems

D.2.1 Chua's oscillator

The circuit diagram of Chua's oscillator [109, 101] is shown in Fig. D.6. Chua's oscillator is a third order autonomous system which can be easily realized in electronic form [110] and exhibits a wide variety of nonlinear and chaotic phenomena. The state equations are given by:

$$\begin{aligned} \frac{dv_1}{dt} &= \frac{1}{C_1}\left(\frac{v_2 - v_1}{R} - f(v_1)\right) \\ \frac{dv_2}{dt} &= \frac{1}{C_2}\left(\frac{v_1 - v_2}{R} + i_3\right) \\ \frac{di_3}{dt} &= -\frac{1}{L}(v_2 + R_L i_3) \end{aligned} \tag{D.8}$$

The dimensionless state equations are given by:

¶Since the nonlinear resistor has a passive and monotone v-i characteristic, the linear resistor must be active for the system to become chaotic (and exhibit sensitive dependence on initial conditions) as otherwise the system will have a unique steady state solution [24].

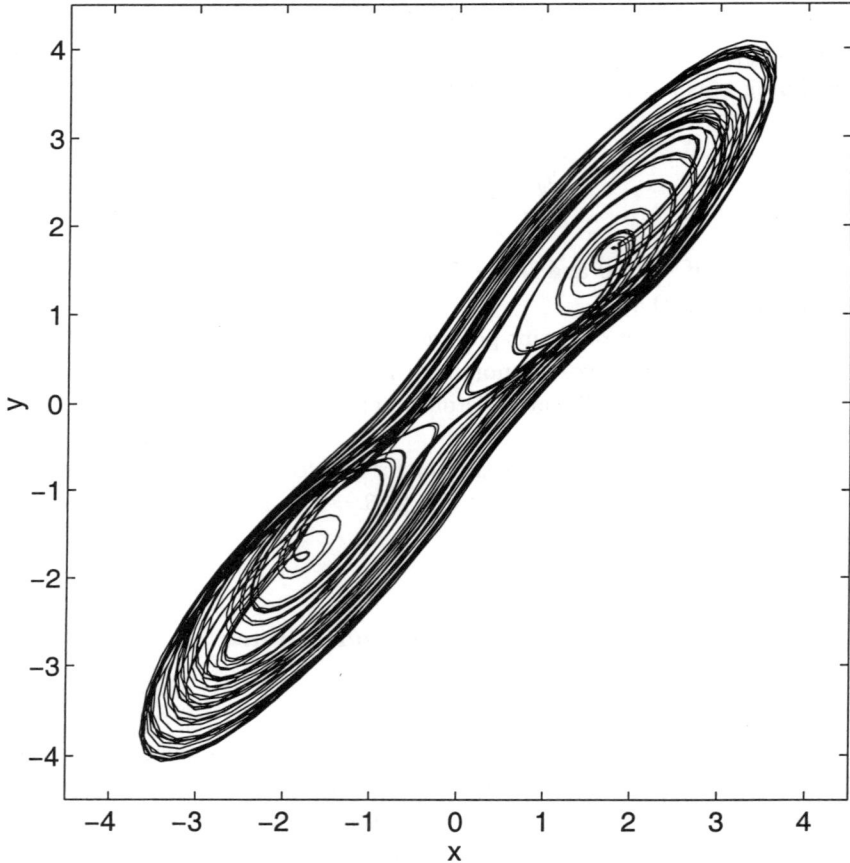

Fig. D.5 Chaotic attractor for system (D.7) in the x-y plane. The parameters are $a = -1.27$, $b = -0.68$, $\omega = 0.5$, $A = 0.2$, $\beta = 1.4$, and $k = -1$.

$$\begin{aligned} \frac{dx}{dt} &= k\alpha(y - x - f(x)) \\ \frac{dy}{dt} &= k(x - y + z) \\ \frac{dz}{dt} &= k(-\beta y - \gamma z) \end{aligned} \quad (D.9)$$

A typical attractor in the x-y plane is shown in Fig. D.7. One important property of Eq. (D.9) is that by selecting appropriate parameters, Chua's oscillator is linearly conjugate to a large class of Lur'e systems [55, 111,

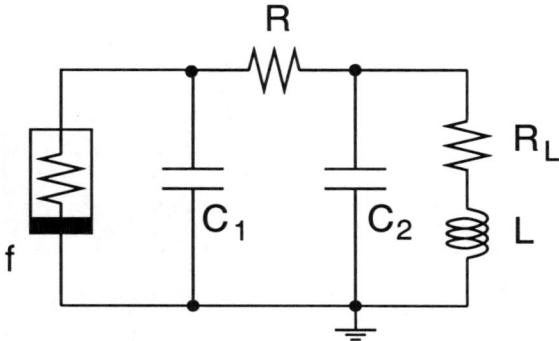

Fig. D.6 Chua's oscillator circuit.

112]:

Theorem D.1 *Chua's oscillator is topologically conjugate via an affine transformation to almost every member of the class of 3-rd order differential equations of the form $\dot{x} = Ax + f(w^T x)b + c$ where A is a 3 by 3 matrix, f is a real-valued function and w, b, c are 3×1 vectors.*

D.2.2 *Piecewise-linear Rössler system*

The piecewise-linear Rössler system [113] is defined by the following state equations:

$$\begin{aligned} \dot{x} &= -k(ax + by + cz) \\ \dot{y} &= k(x + fy) \\ \dot{z} &= -k(-hg(x) + z) \end{aligned} \quad \text{(D.10)}$$

where

$$g(x) = \begin{cases} x - 3 & \text{if } x > 3 \\ 0 & \text{otherwise} \end{cases}$$

A plot of the attractor for system (D.10) in the x-y plane is shown in Fig. D.8 for the parameters $k = 1$, $a = 0.05$, $b = 0.5$, $c = 1$, $f = 0.113$, and $h = 15$.

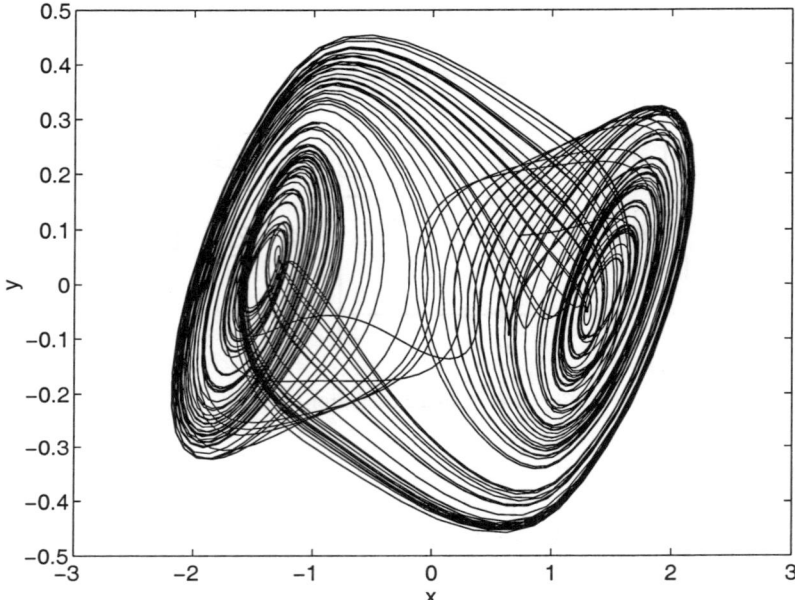

Fig. D.7 Attractor from Chua's oscillator in the x-y plane. The parameters are: $\alpha = 9$, $\beta = 14$, $\gamma = 0.01$, $k = 1$, $f(x) = -0.714x - 0.213\left(\|x+1\| - \|x-1\|\right)$.

D.2.3 *Hyperchaotic electronic circuit*

The state equations of the hyperchaotic circuit studied in [114] are given by:

$$\begin{aligned}
\frac{dv_1}{dt} &= \frac{1}{C_1}\left(f(v_2 - v_1) - i_1\right) \\
\frac{dv_2}{dt} &= \frac{1}{C_2}\left(-f(v_2 - v_1) - i_2\right) \\
\frac{di_1}{dt} &= \frac{1}{L_1}\left(v_1 + Ri_1\right) \\
\frac{di_2}{dt} &= \frac{1}{L_2}\left(v_2\right)
\end{aligned} \quad (\text{D.11})$$

A circuit diagram of this circuit is shown in Fig. D.9. The parameters used are: $C_1 = 0.5$, $C_2 = 0.05$, $L_1 = 1$, $L_2 = 2/3$, $R = 1$. The nonlinear function f is given by Eq. D.4 where $a = -0.2$ and $b = 3$. A chaotic attractor for these parameters in the v_2-i_1 plane is shown in Fig. D.10.

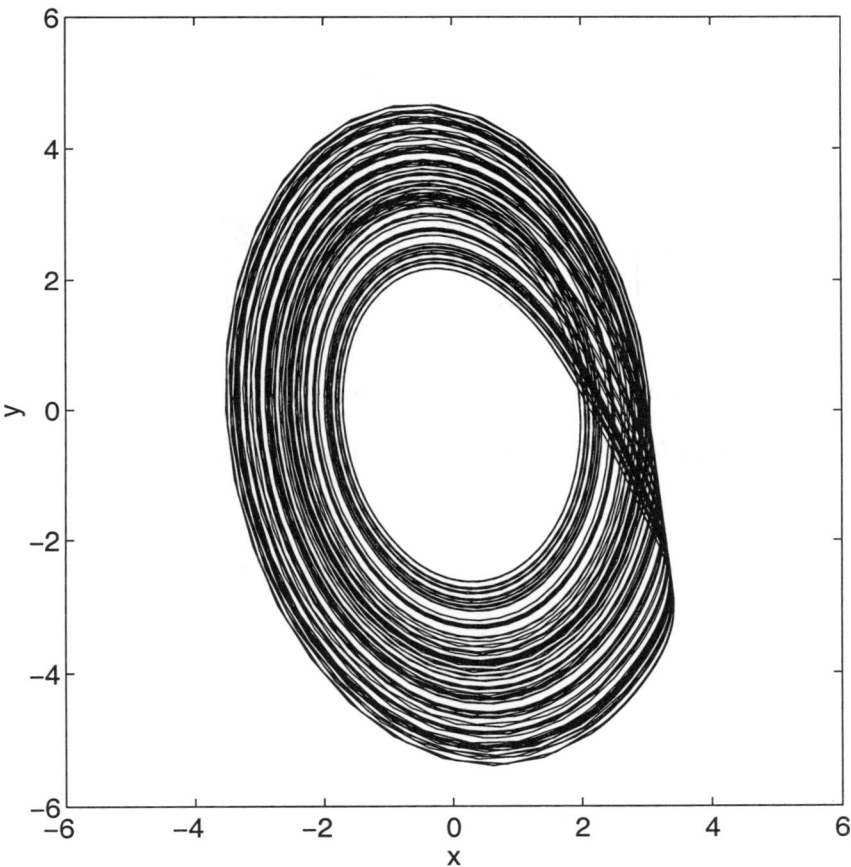

Fig. D.8 Chaotic attractor for the piecewise-linear Rössler system (D.10) in the x-y plane. The parameters are $k = 1$, $a = 0.05$, $b = 0.5$, $c = 1$, $f = 0.113$, and $h = 15$.

D.2.4 Hyperchaotic Rössler system

The state equations for the hyperchaotic Rössler system are given by [115]:

$$\begin{aligned} \frac{dx}{dt} &= -y - z \\ \frac{dy}{dt} &= x + 0.25y + w \\ \frac{dz}{dt} &= 3 + xz \\ \frac{dw}{dt} &= -0.5z + 0.05w \end{aligned} \tag{D.12}$$

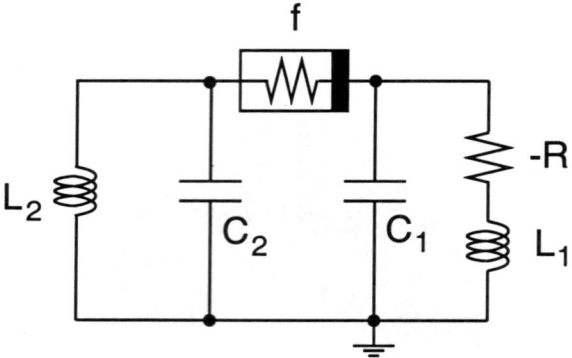

Fig. D.9 Circuit diagram of hyperchaotic electronic circuit.

A chaotic attractor of this system in the y-w plane is shown in Fig. D.11.

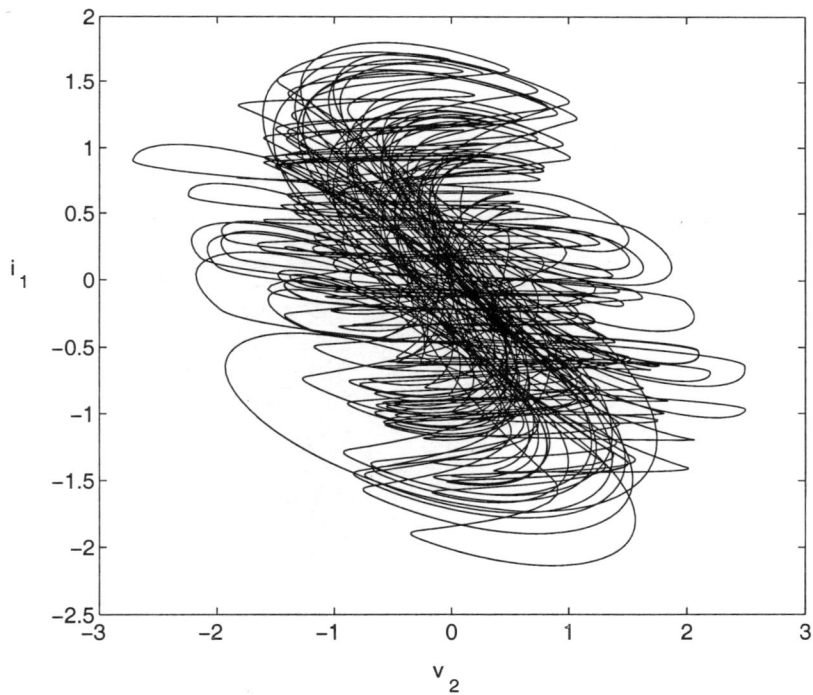

Fig. D.10 Attractor from a hyperchaotic electronic circuit in the v_2-i_1 plane. The parameters are $C_1 = 0.5$, $C_2 = 0.05$, $L_1 = 1$, $L_2 = 2/3$, $R = 1$, $a = -0.2$ and $b = 3$.

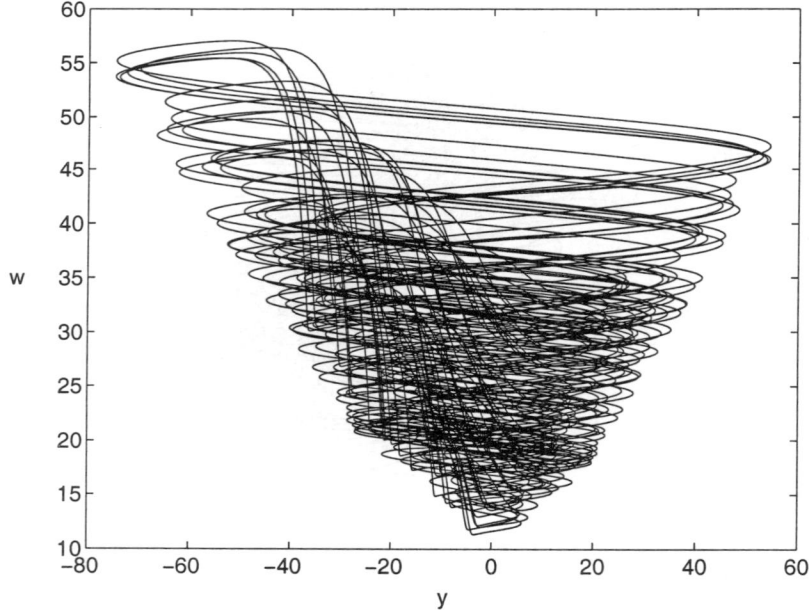

Fig. D.11 Attractor from the hyperchaotic Rössler system in the y-w plane.

Bibliography

[1] B. V. der Pol and J. V. der Mark, "Frequency demultiplication," *Nature*, vol. 120, pp. 363–364, Sept. 1927.

[2] M. P. Kennedy and L. O. Chua, "Van der pol and chaos," *IEEE Transactions on Circuits and Systems*, vol. 33, no. 10, pp. 974–980, 1986.

[3] L. M. Pecora and T. L. Carroll, "Synchronization in chaotic systems," *Physical Review Letters*, vol. 64, pp. 821–824, Feb. 1990.

[4] L. M. Pecora and T. L. Carroll, "Driving systems with chaotic signals," *Physical Review A*, vol. 44, pp. 2374–2383, Aug. 1991.

[5] V. S. Anishchenko, M. A. Safonova, and L. O. Chua, "Stochastic resonance in Chua's circuit," *International Journal of Bifurcation and Chaos*, vol. 2, no. 2, pp. 397–401, 1992.

[6] L. O. Chua, C. A. Desoer, and E. S. Kuh, *Linear and Nonlinear Circuits*. New York: McGraw-Hill, 1987.

[7] R. He and P. G. Vaidya, "Analysis and synthesis of synchronous periodic and chaotic systems," *Physical Review A*, vol. 46, pp. 7387–7392, Dec. 1992.

[8] C. W. Wu and L. O. Chua, "A unified framework for synchronization and control of dynamical systems," *International Journal of Bifurcation and Chaos*, vol. 4, no. 4, pp. 979–998, 1994.

[9] L. Kocarev and U. Parlitz, "General approach for chaotic synchronization with applications to communications," *Physical Review Letters*, vol. 74, no. 25, pp. 5028–5031, 1995.

[10] J. A. K. Suykens, P. F. Curran, J. Vandewalle, and L. O. Chua, "Robust nonlinear H_∞ synchronization of chaotic lur'e systems," *IEEE Transactions on Circuits and Systems–I: Fundamental Theory and Applications*, vol. 44, no. 10, pp. 891–904, 1997.

[11] G. Grassi and S. Mascolo, "Nonlinear observer design to synchronize hyperchaotic systems via a scalar signal," *IEEE Transactions on Circuits and Systems–I: Fundamental Theory and Applications*, vol. 44, no. 10, pp. 1011–1014, 1997.

[12] H. Nijmeijer and I. M. Y. Mareels, "An observer looks at synchronization," *IEEE Transactions on Circuits and Systems–I: Fundamental Theory and Applications*, vol. 44, no. 10, pp. 882–890, 1997.

[13] C. W. Wu and L. O. Chua, "Synchronization in an array of linearly coupled dynamical systems," *IEEE Transactions on Circuits and Systems–I: Fundamental Theory and Applications*, vol. 42, no. 8, pp. 430–447, 1995.

[14] A. V. Oppenheim, G. W. Wornell, S. H. Isabelle, and K. M. Cuomo, "Signal processing in the context of chaotic signals," *Proc 1992 IEEE ICASSP*, vol. IV, pp. 117–120, 1992.

[15] L. Kocarev, K. S. Halle, K. Eckert, L. O. Chua, and U. Parlitz, "Experimental demonstration of secure communications via chaotic synchronization," *International Journal of Bifurcation and Chaos*, vol. 2, no. 3, pp. 709–713, 1992.

[16] G. Chen and X. Dong, "On feedback control of chaotic continuous-time systems," *IEEE Transactions on Circuits and Systems–I: Fundamental Theory and Applications*, vol. 40, no. 9, pp. 591–601, 1993.

[17] G. Malescio, "Synchronization of chaotic systems by continuous control," *Physical Review E*, vol. 53, no. 3, pp. 2949–2952, 1996.

[18] P. F. Curran, J. A. K. Suykens, and L. O. Chua, "Absolute stability theory and master-slave synchronization," *International Journal of Bifurcation and Chaos*, vol. 7, no. 12, pp. 2891–2896, 1997.

[19] L. O. Chua and D. N. Green, "A qualitative analysis of the behavior of dynamic nonlinear networks: Stability of autonomous networks," *IEEE Transactions on Circuits and Systems*, vol. 23, no. 6, pp. 355–379, 1976.

[20] L. O. Chua and D. N. Green, "Graph-theoretical properties of dynamic nonlinear networks," Tech. Rep. Memo ERL-M507, College of Engineering, University of California, Berkeley, 1975.

[21] M. Vidyasagar, *Nonlinear Systems Analysis*. New Jersey: Prentice-Hall, 2nd ed., 1993.

[22] R. Rajamani, "Observers for Lipschitz nonlinear systems," *IEEE Transactions on Automatic Control*, vol. 43, no. 3, pp. 397–401, 1998.

[23] X. F. Wang, Z. Q. Wang, and G. Chen, "A new criterion for synchronization of coupled chaotic oscillators with application to Chua's circuits," *International Journal of Bifurcation and Chaos*, vol. 9, no. 6, pp. 1169–1174, 1999.

[24] L. O. Chua and D. N. Green, "A qualitative analysis of the behavior of dynamic nonlinear networks: Steady-state solutions of nonautonomous networks," *IEEE Transactions on Circuits and Systems*, vol. 23, no. 9, pp. 530–550, 1976.

[25] L. O. Chua, "Dynamic nonlinear networks: State-of-the-art," *IEEE Transactions on Circuits and Systems*, vol. 27, pp. 1059–1087, Nov. 1980.

[26] C. W. Wu, "Qualitative analysis of dynamic circuits," in *Wiley Encyclopedia of Electrical and Electronics Engineering* (J. G. Webster, ed.), vol. 17, pp. 500–510, John G. Wiley & Sons, 1999.

[27] M. J. Hasler and P. Verburgh, "On the uniqueness of the steady state for nonlinear circuits with time-dependent sources," *IEEE Transactions on Circuits and Systems*, vol. 31, no. 8, pp. 702–713, 1984.

[28] K. S. Halle, C. W. Wu, M. Itoh, and L. O. Chua, "Spread spectrum communication through modulation of chaos," *International Journal of Bifurcation and Chaos*, vol. 3, no. 2, pp. 223–239, 1993.

[29] C. W. Wu and L. O. Chua, "A simple way to synchronize chaotic systems with applications to secure communication systems," *International Journal of Bifurcation and Chaos*, vol. 3, no. 6, pp. 1619–1627, 1993.

[30] U. Feldmann, M. Hasler, and W. Schwarz, "Communication by chaotic signals: the inverse system approach," in *Proceedings 1995 IEEE International Symposium on Circuits and Systems*, vol. 1, pp. 680–683, 1995.

[31] G. Kolumbán, M. P. Kennedy, and L. O. Chua, "The role of synchronization in digital communications using chaos-part i:fundamentals of digital communications," *IEEE Transactions on Circuits and Systems–I: Fundamental Theory and Applications*, vol. 44, no. 10, pp. 927–936, 1997.

[32] "Special issue on noncoherent chaotic communications." IEEE Transactions on Circuits and Systems–I: Fundamental Theory and Applications, vol. 47, no. 12, 2000.

[33] M. P. Kennedy, G. Kolumbán, G. Kis, and Z. Jákó, "Performance evaluation of FM-DCSK modulation in multipath environments," *IEEE Transactions on Circuits and Systems–I: Fundamental Theory and Applications*, vol. 47, no. 12, pp. 1702–1711, 2000.

[34] K. Murali, M. Lakshmanan, and L. O. Chua, "Controlling and synchronization of chaos in the simplest dissipative non-autonomous circuit," *International Journal of Bifurcation and Chaos*, vol. 4, no. 6, 1994.

[35] T. L. Carroll and L. M. Pecora, "Synchronizing nonautonomous chaotic circuits," *IEEE Transactions on Circuits and Systems–II: Analog and Digital Signal Processing*, vol. 40, no. 10, pp. 646–650, 1993.

[36] C. W. Wu, G. Q. Zhong, and L. O. Chua, "Synchronizing nonautonomous chaotic systems without phase-locking," *Journal of Circuits, Systems, and Computers*, vol. 6, no. 3, pp. 227–241, 1996.

[37] K. S. Halle, C. W. Wu, M. Itoh, and L. O. Chua, "Spread spectrum communication through modulation of chaos," *International Journal of Bifurcation and Chaos*, vol. 3, no. 2, pp. 469–477, 1993.

[38] U. Parlitz, L. O. Chua, L. Kocarev, K. S. Halle, and A. Shang, "Transmission of digital signals by chaotic synchronization," *International Journal of Bifurcation and Chaos*, vol. 2, no. 4, pp. 973–977, 1992.

[39] K. M. Cuomo and A. V. Oppenheim, "Circuit implementation of synchronized chaos with applications to communications," *Physical Review Letters*, vol. 71, no. 1, pp. 65–68, 1993.

[40] K. M. Cuomo and A. V. Oppenheim, "Chaotic signals and systems for communications," in *Proceedings of 1993 IEEE ICASSP III*, pp. 137–140, 1993.

[41] H. Dedieu, M. P. Kennedy, and M. Hasler, "Chaos shift keying: Modulation and demodulation of a chaotic carrier using self-synchronizing Chua's circuits," *IEEE Transactions on Circuits and Systems–II: Analog and Digital Signal Processing*, vol. 40, no. 10, pp. 634–642, 1993.

[42] J. H. Peng, E. J. Ding, M. Ding, and W. Yang, "Synchronizing hyperchaos with a scalar transmitted signal," *Physical Review Letters*, vol. 76, pp. 904–907, Feb. 1996.

[43] A. Tamasevicius and A. Cenys, "Synchronizing hyperchaos with a single variable," *Physical Review E*, vol. 55, pp. 297–299, Jan. 1997.

[44] M. Itoh, C. W. Wu, and L. O. Chua, "Communication systems via chaotic signals from a reconstruction viewpoint," *International Journal of Bifurcation and Chaos*, vol. 7, pp. 275–286, Feb. 1997.

[45] C.-T. Chen, *Linear System Theory and Design*. New York: Holt, Rinehart and Winston, 1984.

[46] C. Tresser, P. A. Worfolk, and H. Bass, "Master-slave synchronization from the point of view of global dynamics," *Chaos*, vol. 5, no. 4, pp. 693–699, 1995.

[47] B. A. Huberman and E. Lumer, "Dynamics of adaptive systems," *IEEE Transactions on Circuits and Systems*, vol. 37, no. 4, pp. 547–550, 1990.

[48] S. Sinha, R. Ramaswamy, and J. S. Rao, "Adaptive control in nonlinear dynamics," *Physica D*, vol. 43, pp. 118–128, 1990.

[49] J. K. John and R. E. Amritkar, "Synchronization by feedback and adaptive control," *International Journal of Bifurcation and Chaos*, vol. 4, no. 6, pp. 1687–1695, 1994.

[50] P. Celka, "Synchronization of chaotic systems through parameter adaptation," in *IEEE International Symposium of Circuits and Systems Proceedings*, vol. 1, pp. 692–695, Apr. 1995.

[51] L. O. Chua, T. Yang, G.-Q. Zhong, and C. W. Wu, "Adaptive synchronization of Chua's oscillators," *International Journal of Bifurcation and Chaos*, vol. 6, no. 1, pp. 189–201, 1996.

[52] C. W. Wu, T. Yang, and L. O. Chua, "On adaptive synchronization and control of nonlinear dynamical systems," *International Journal of Bifurcation and Chaos*, vol. 6, no. 2, 1996.

[53] C. W. Wu and L. O. Chua, "On a variation of the Huberman-Lumer adaptive scheme," *International Journal of Bifurcation and Chaos*, vol. 6, no. 7, pp. 1397–1407, 1996.

[54] P. M. Clarkson, *Optimal and Adaptive Signal Processing*. CRC Press, 1993.

[55] C. W. Wu and L. O. Chua, "On the generality of the unfolded Chua's circuit," *International Journal of Bifurcation and Chaos*, vol. 6, no. 5, pp. 801–832, 1996.

[56] R. Marino and P. Tomei, "Global adaptive observers for nonlinear systems via filtered transformations," *IEEE Transactions on Automatic Control*, vol. 37, no. 8, pp. 1239–1245, 1992.

[57] U. Parlitz and L. Kocarev, "Multichannel communication using autosyn-

chronization," *International Journal of Bifurcation and Chaos*, vol. 6, no. 3, pp. 581–588, 1996.

[58] E. Ott, *Chaos in dynamical systems*. Cambridge University Press, 1993.

[59] H. Nijmeijer and A. V. D. Schaft, *Nonlinear Dynamical Control Systems*. Springer Verlag, 1990.

[60] S. Sastry, *Nonlinear Systems: Analysis, Stability and Control*. Springer, 1999.

[61] M. di Bernardo, "An adaptive approach to the control and synchronization of continuous-time chaotic systems," *International Journal of Bifurcation and Chaos*, vol. 6, no. 3, pp. 557–568, 1996.

[62] A. L. Fradkov, H. Nijmeijer, and A. Y. Pogromsky, "Adaptive observer-based synchronization," in *Controlling Chaos and Bifurcations in Engineering Systems* (G. Chen, ed.), ch. 19, pp. 417–438, CRC Press, 1999.

[63] "Special issue on chaos synchronization and control: theory and applications." IEEE Transactions on Circuits and Systems–I: Fundamental Theory and Applications, vol. 44, no. 10, 1997.

[64] "Special issue on control and synchronization of chaos." Chaos, vol. 7, no. 4, 1997.

[65] C. W. Wu and L. O. Chua, "Application of Kronecker products to the analysis of systems with uniform linear coupling," *IEEE Transactions on Circuits and Systems–I: Fundamental Theory and Applications*, vol. 42, no. 10, pp. 775–778, 1995.

[66] F. R. Gantmacher, *The Theory of Matrices*, vol. 1. New York: Chelsea Publishing Company, 1960.

[67] L. Vandenberghe and S. Boyd, "A polynomial-time algorithm for determining quadratic Lyapunov functions for nonlinear systems," in *Proceedings of the European Conference on Circuit Theory and Design*, pp. 1065–1068, 1993.

[68] R. Fletcher, *Practical methods of optimization*. John Wiley & Sons, 2nd ed., 1987.

[69] Y. Nesterov and A. Nemirovskii, *Interior-Point Polynomial Algorithms in Convex Programming*, vol. 13 of *SIAM Studies in Applied Mathematics*. Society for Industrial and Applied Mathematics, 1994.

[70] C. W. Wu, "Synchronization in arrays of chaotic circuits coupled via hypergraphs: static and dynamic coupling," in *Proceedings of the 1998 IEEE Int. Symp. Circ. Syst.*, vol. 3, pp. III-287–290, IEEE, 1998.

[71] C. W. Wu, "Synchronization in arrays of coupled nonlinear systems: passivity, circle criterion and observer design," in *IEEE International Symposium on Circuits and Systems*, pp. III-692–695, 2001.

[72] C. W. Wu, "Synchronization in arrays of coupled nonlinear systems: Passivity, circle criterion and observer design," *IEEE Transactions on Circuits and Systems–I: Fundamental Theory and Applications*, vol. 48, no. 10, pp. 1257–1261, 2001.

[73] K. S. Fink, G. Johnson, T. Carroll, D. Mar, and L. Pecora, "Three cou-

pled oscillators as a universal probe of synchronization stability in coupled oscillator arrays," *Physical Review E*, vol. 61, pp. 5080–5090, May 2000.

[74] K. Kaneko, "Overview of coupled map lattices," *CHAOS*, vol. 2, no. 3, pp. 279–282, 1992.

[75] K. Kaneko, "Chaotic but regular posi-nega switch among coded attractors by cluster-size variation," *Physical Review Letters*, vol. 63, pp. 219–223, July 1989.

[76] H. Chaté and P. Manneville, "Collective behaviors in coupled map lattices with local and nonlocal connections," *Chaos*, vol. 2, no. 3, pp. 307–313, 1992.

[77] J. Friedman, J. Kahn, and E. Szemerédi, "On the second eigenvalue in random regular graphs," in *Proceedings of the 21st Annual ACM Symposium on Theory of Computing*, pp. 587–598, 1989.

[78] L. Pivka, A. L. Zheleznyak, C. W. Wu, and L. O. Chua, "On the generation of scroll waves in a three-dimensional discrete active medium," *International Journal of Bifurcation and Chaos*, vol. 5, no. 1, pp. 313–320, 1995.

[79] V. Perez-Muñuzuri, A. Muñuzuri, M. Gomez-Gesteira, V. Perez-Villar, L. Pivka, and L. Chua, "Nonlinear waves, patterns and spatio-temporal chaos in cellular neural networks," *Phil. Trans. Roy. Soc. London A*, vol. 353, pp. 101–113, 1995.

[80] C. W. Wu, "Synchronization in arrays of chaotic circuits coupled via dynamic coupling elements," *International Journal of Bifurcation and Chaos*, vol. 10, no. 4, pp. 819–827, 2000.

[81] R. Merris, "Laplacian matrices of graphs: a survey," *Linear algebra and its applications*, vol. 197–198, pp. 143–176, 1994.

[82] F. R. K. Chung, *Spectral Graph Theory*. CBMS regional conference series in mathematics, no. 92, American Mathematical Society, 1997.

[83] C. W. Wu and L. O. Chua, "Application of graph theory to the synchronization in an array of coupled nonlinear oscillators," *IEEE Transactions on Circuits and Systems–I: Fundamental Theory and Applications*, vol. 42, no. 8, pp. 494–497, 1995.

[84] M. Fiedler, "Algebraic connectivity of graphs," *Czechoslovak Mathematical Journal*, vol. 23, no. 98, pp. 298–305, 1973.

[85] B. Mohar, "Isoperimetric numbers of graphs," *Journal of Combinatorial Theory, Series B*, vol. 47, pp. 274–291, 1989.

[86] B. Mohar, "Eigenvalues, diameter, and mean distance in graphs," *Graphs and Combinatorics*, vol. 7, pp. 53–64, 1991.

[87] P. M. Gade, "Synchronization in coupled map lattices with random nonlocal connectivity," *Physical Review E*, vol. 54, no. 1, pp. 64–70, 1996.

[88] B. Bollobás, "The isoperimetric number of random regular graphs," *European Journal of Combinatorics*, vol. 9, pp. 241–244, 1988.

[89] D. J. Watts and S. H. Strogatz, "Collective dynamics of 'small-world' networks," *Nature*, vol. 393, pp. 440–442, 1998.

[90] G. W. Stewart and J.-G. Sun, *Matrix Perturbation Theory*. Academic Press, 1990.

[91] R. Grone and R. Merris, "The Laplacian spectrum of a graph II," *SIAM Journal of Discrete Mathematics*, vol. 7, no. 2, pp. 221–229, 1994.

[92] A. B. Carlson and D. G. Gisser, *Electrical Engineering: Concepts and Applications*. Addison-Wesley Publishing Company, 1981.

[93] C. W. Wu, "Graph coloring via synchronization of coupled oscillators," *IEEE Transactions on Circuits and Systems–I: Fundamental Theory and Applications*, vol. 45, no. 9, pp. 974–978, 1998.

[94] L. Kučera, *Expected behaviour of graph colouring algorithms*, vol. 56 of *Lecture Notes in Computer Science*, pp. 477–483. Springer-Verlag, 1977.

[95] A. D. Petford and D. J. A. Welsh, "A randomised 3-colouring algorithm," *Discrete Mathematics*, vol. 74, pp. 253–261, 1989.

[96] A. Vince, "Star chromatic number," *Journal of Graph Theory*, vol. 12, no. 4, pp. 551–559, 1988.

[97] H. Fujisaka and T. Yamada, "Stability theory of synchronized motion in coupled-oscillator systems," *Progress of Theoretical Physics*, vol. 69, pp. 32–47, Jan. 1983.

[98] L. M. Pecora and T. L. Carroll, "Master stability functions for synchronized chaos in arrays of oscillators," in *Proceedings of the 1998 IEEE Int. Symp. Circ. Syst.*, vol. 4, pp. IV-562–567, IEEE, 1998.

[99] A. S. Dmitriev, M. Shirokov, and S. O. Starkov, "Chaotic synchronization in ensembles of coupled maps," *IEEE Transactions on Circuits and Systems–I: Fundamental Theory and Applications*, vol. 44, no. 10, pp. 918–926, 1997.

[100] C. W. Wu, "Simple three oscillator universal probes for determining synchronization stability in coupled arrays of oscillators," in *IEEE International Symposium on Circuits and Systems*, pp. III-261–264, 2001.

[101] L. O. Chua, C. W. Wu, A. Huang, and G. Q. Zhong, "A universal circuit for studying and generating chaos, part I: Routes to chaos," *IEEE Transactions on Circuits and Systems–I: Fundamental Theory and Applications*, vol. 40, no. 10, pp. 732–744, 1993. Special Issue on Chaos in Electronic Circuits, Part A.

[102] G. H. Golub and C. E. V. Loan, *Matrix Computations*. Johns Hopkins University Press, second ed., 1989.

[103] H. Minc, *Nonnegative Matrices*. New York: John Wiley & Sons, 1988.

[104] M. Bolla, "Spectra, euclidean representations and clusterings of hypergraphs," *Discrete Mathematics*, vol. 117, pp. 19–39, 1993.

[105] T. Yoshizawa, *Stability Theory by Liapunov's Second Method*. Tokyo, Japan: The mathematical society of Japan, 1966.

[106] V. Lakshmikantham and X. Z. Liu, *Stability Analysis In Terms of Two Measures*. Singapore: World Scientific, 1993.

[107] T. Parker and L. O. Chua, *Practical Numerical Algorithms for Chaotic Systems*. Springer-Verlag, 1989.

[108] K. Murali, M. Lakshmanan, and L. O. Chua, "The simplest dissipative nonautonomous chaotic circuit," *IEEE Transactions on Circuits and Systems–I: Fundamental Theory and Applications*, vol. 41, no. 6, pp. 462–463, 1994.

[109] L. O. Chua, "The genesis of Chua's circuit," *Archiv für Elektronik und Übertragungstechnik*, vol. 46, no. 4, pp. 250–257, 1992.

[110] M. P. Kennedy, "Robust op amp realization of Chua's circuit," *Frequenz*, vol. 46, no. 3–4, pp. 66–80, 1992.

[111] C. W. Wu and L. O. Chua, "On linear topological conjugacy of Lur'e systems," *IEEE Transactions on Circuits and Systems–I: Fundamental Theory and Applications*, vol. 43, no. 2, pp. 158–161, 1996.

[112] L. Kocarev and T. D. Stojanovski, "Linear conjugacy of vector fields in Lur'e form," *IEEE Transactions on Circuits and Systems–I: Fundamental Theory and Applications*, vol. 43, no. 9, pp. 782–785, 1996.

[113] T. L. Carroll, "A simple circuit for demonstrating regular and synchronized chaos," *American Journal of Physics*, vol. 63, no. 4, pp. 377–379, 1995.

[114] T. Matsumoto, L. O. Chua, and K. Kobayashi, "Hyperchaos: Laboratory experiment and numerical confirmation," *IEEE Transactions on Circuits and Systems*, vol. 33, no. 11, pp. 1143–1147, 1986.

[115] O. E. Rossler, "An equation for hyperchaos," *Physics Letters*, vol. 71A, pp. 155–157, Apr. 1979.

Index

M_1, 134
M_2, 134, 150
W, 57, 135
W_i, 57, 135, 149

algebraic connectivity, 61, 88, 89, 91, 92, 95, 104, 109, 141
autonomous, 18
autonomous system, 70

Cholesky decomposition, 71
Chua's oscillator, 157
circle criterion, 9
circulant, 101, 102
condition number, 70, 148
connectivity graph, 52
connectivity hypergraph, 52, 78
connectivity matrix, 101
controllable, 10, 29
coupled map lattices, 70
coupling
    additive, 6
    bidirectional, 20
    dynamic, 77
    master-slave, 5
    mutual, 20
    reciprocal, 75, 108
    static, 55
    unidirectional, 5
Courant-Fischer theorem, 107, 135, 136

decreasing functions, 8
    V-uniformly, 8
    uniformly, 8
diagonalizable, 120, 131
dual coupled system, 80

eigenvalue assignment, 27

Gershgorin's circle criterion, 63, 135
graph
    chromatic number, 142
    diameter, 104, 142
    distance, 142
    edge connectivity, 104, 142
    edge density, 105
    isoperimetric number, 104, 106, 142
    mean distance, 104, 142

Hurwitz, 131
hypergraph, 141
    dual, 80, 141

increasing functions, 8
    V-uniformly, 8
    strictly, 8
    uniformly, 8
irreducible matrix, 132

Kirchhoff's theorem, 137
Kronecker product, 55

Laplacian matrix, 61, 99, 141
Lipschitz continuous, 9, 11
LMS
    Clipped, 36
    Pilot, 36
Lur'e systems, 158
Lyapunov
    direct method, 145
    exponents, 119, 150

matrix
    normal, 58, 132
minimal realization, 9, 131

negative definite, 131
negative definite operator, 66
nonautonomous, 18

observable, 10, 29
observer, 7, 27
odd-symmetric, 38, 109

passive
    strictly, 35
passivity, 9
Pecora-Carroll decomposition, 6, 97
Perron-Frobenius theory, 135
phase-locking, 18
positive definite, 131
pseudo-inverse, 82

Rössler system, 159
reducible matrix, 132
relative degree, 16

Schwarz's inequality, 9
sensitive dependence on initial
    conditions, 3
Signed Error algorithm, 36
Signed Regressor algorithm, 36
stability
    absolute, 8
    asymptotical, 145
star chromatic number, 116
strictly positive real, 9, 132
Substitution Theorem, 77, 84
Sylvester's inequality, 137, 139
synchronizable, 28
synchronization
    adaptive, 33
    clustered, 82
synchronization subspace, 99, 120

tensor product, 55
Thévenin-Norton transformation, 84

unique steady state, 11

Vandemonde matrix, 32